A New Foundation of Physical Theories

Günther Ludwig   Gérald Thurler

# A New Foundation of Physical Theories

 Springer

Professor Dr. Günther Ludwig
Sperberweg 11
34043 Marburg/Lahn
Germany

Dr. Gérald Thurler
Rue Baulacre 30
1202 Genève
Switzerland

Library of Congress Control Number: 2006922616

ISBN-10  3-540-30832-6 Springer Berlin Heidelberg New York
ISBN-13  978-3-540-30832-4 Springer Berlin Heidelberg New York

This work is subject to copyright. All rights are reserved, whether the whole or part of the material is concerned, specifically the rights of translation, reprinting, reuse of illustrations, recitation, broadcasting, reproduction on microfilm or in any other way, and storage in data banks. Duplication of this publication or parts thereof is permitted only under the provisions of the German Copyright Law of September 9, 1965, in its current version, and permission for use must always be obtained from Springer. Violations are liable for prosecution under the German Copyright Law.

Springer is a part of Springer Science+Business Media
springer.com
© Springer-Verlag Berlin Heidelberg 2006
Printed in The Netherlands

The use of general descriptive names, registered names, trademarks, etc. in this publication does not imply, even in the absence of a specific statement, that such names are exempt from the relevant protective laws and regulations and therefore free for general use.

Typesetting: by the author and techbooks using a Springer LATEX macro package
Cover design: *design & production* GmbH, Heidelberg

Printed on acid-free paper    SPIN: 11548744    55/techbooks    5 4 3 2 1 0

# Foreword

(translation)

I was interested by the development of a new edition of the book [1]

"Die Grundstrukturen einer physikalischen Theorie."

This has been possible, in spite of my old age, thanks to the contributions of Dr. G. Thurler. Without his indefatigable support and his essential and fundamental propositions, this new edition would not have been possible.

The new edition clarifies and formulates more precisely the fundamental ideas of physical theories in order to avoid as much as possible any ambiguities. One begins theoretical physics with concepts that can be explained without theories. Later, one introduces other concepts by theories known as "pre-theories." Thus it does not make sense to introduce concepts such as "state" without a pre-theory.

The field of physics is thus determined by the basic concepts introduced without the use of pre-theories. Also, it does not make sense to speak about the position and speed of an electron at a fixed time.

"Reality" is not however only the reality which is described by physical concepts. Thus, for example, colors, tones, joy, hate, and love are not physical concepts.

But the demarcation of the physical concepts, and thus the demarcation of the field of physics makes it possible to know more clearly, and thus to describe more clearly in the future, the structure of reality beyond the domain of physics. The field of life and not that of death should be the goal of mankind. Thus, I hope that this book can also become another small step for life.

Marburg  *Günther Ludwig*
October 2005

# Vorwort

Ich war daran interessiert, bald eine neue Auflage des Buches

„Die Grundstrukturen einer physikalischen Theorie"

zu entwerfen ( [1]). Daß dies trotz meines hohen Alters möglich wurde, habe ich Herrn Dr. G. Thurler zu verdanken. Ohne seine unermüdliche Hilfe und seine wesentlichen Vorschläge auch in haltlicher Art, wäre die Neuauflage nie Zustande gekommen.

Diese Neuauflage soll die Grundsätzlichen Ideen klären und präziser formulieren, um möglichst jede Fehlentwicklung physikalischer Theorien zu vermeiden. Dazu gehört, daß man die theoretische Physik nur mit Begriffen anfängt, die ohne jede Theorie erklärt werden können. Später führt man dann mit Hilfe von Theorien (sogenannten Vortheorien) weitere Begriffe ein. So macht es keinen Sinn, den Begriff „Zustand" ohne eine Vortheorie einzuführen.

Der Umfang der Physik ist damit bestimmt durch die ohne Vortheorien eingeführten Grundbegriffe. Ebenso macht es keinen Sinn, von Ort und Geschwindigkeit eines Elektrons zu einer festen Zeit zu sprechen.

Die Wirklichkeit ist aber nicht die allein mit physikalischen Begriffen beschriebene Wirklichkeit. So sind z.B. Farben, Töne, Freude, Haß und Liebe keine physikalischen Begriffe.

Aber die saubere Abgrenzung der physikalischen Begriffe und damit die saubere Abgrenzung des Bereichs der Physik wird es möglich machen, in der Zukunft auch die Struktur der über den physikalischen Bereich hinausgehenden Wirklichkeit deutlicher zu erfahren und damit auch deutlicher zu beschreiben. Der Bereich des Lebens und nicht der des Todes ist das Ziel des Menschen. So hoffe ich, daß auch dieses Buch eine kleiner Schritt zum Leben werden kann.

Marburg *Günther Ludwig*
Oktober 2005

# Preface

This book is a revision and expansion of the concept of a physical theory as developed in [1].

In this book, we introduce the following:

- A concept of basic language; a descriptive language of simple form in which it is possible to formulate recorded facts. The semantics of this basic language make it possible to clarify the links between linguistic, conceptual, and real entities of the application domain of a physical theory.
- A new concept of idealization. We know that practically all mathematical theories used in the physical theories can only be approximations of the reality, i.e., that they can be applied to an application domain of a physical theory only under the assumption of allowing for some degree of approximation or degree of inaccuracy.

We propose a review (related to the new concepts introduced above) of the "notion of relations between various physical theories," and of the "process allowing to find new concepts" developed in [1].

The analysis presented here will be less of a description of the current state of physics than a suggestion to modify this state. The authors think that a solution can be found amongst the many difficult problems of physics such as the interpretation of physical theories, the relations between various theories, and the introduction of physical concepts, when the theories are under the form of an axiomatic basis. The analysis presented here does not claim to be definitive. It should, on the contrary, encourage the reader to continue the development of the fundamental ideas of this work. Such a development should contribute to highlight the durable core and growing strength of physical knowledge about the real structures of the world, in addition to the process of the historical development of physics.

If this book was to suggest such a development, it would then have achieved its goal. The authors also encourage the reader to correct any possible faults in the text and are convinced that the correction of such errors will not call into question the fundamental ideas of this work.

## Acknowledgments

The authors wish to express their deep thanks to Natacha Carrara for her careful re-reading and linguistic revision of the English manuscript.

We are also grateful to Wolf Beiglböck for his competent advice and for his assistance in the completion of the book.

Marburg, Genève  *Günther Ludwig*
October 2005  *Gérald Thurler*

# Contents

Intention of the Book .................................................. 1

## Part I  A New Form of Physical Theory

1 Reality .................................................... 11
  1.1 The Structure of Reality ............................... 11
  1.2 The Physical Reality ................................... 12
      1.2.1 The Application Domain of a $PT$ ................. 12
      1.2.2 The Fundamental Domain of a $PT$ ................. 13
      1.2.3 The Reality Domain of a $PT$ ..................... 14
      1.2.4 The Reality Domain of all $PT$s .................. 14
      1.2.5 Remarks .......................................... 15
  1.3 Fairy Tales ............................................ 16

2 Building of a Mathematical Theory ......................... 17
  2.1 Formal Language ........................................ 17
  2.2 Axioms and Proofs ...................................... 19
  2.3 Logics ................................................. 21
  2.4 Set Theory ............................................. 27

3 From Reality to Mathematics ............................... 33
  3.1 Recording Process ...................................... 34
      3.1.1 Basic Language ................................... 34
      3.1.2 Application Domain of a $PT$ ..................... 44
      3.1.3 Recording Rules .................................. 44
      3.1.4 Facts Recorded in the Basic Language ............. 45
  3.2 Mathematization Process ................................ 46
      3.2.1 The Basic Mathematical Theory .................... 46
      3.2.2 The Standard Mathematical Theory ................. 48
      3.2.3 Enrichment of $MT_\Theta$ by $\overline{A}$ ...... 50

|  |  | 3.2.4 | The Finiteness of Physics............................ 52 |
|  | 3.3 | Idealization Process........................................... 53 |
|  |  | 3.3.1 | Transition from $MT_\Theta$ to $MT_\Delta$ ...................... 53 |
|  |  | 3.3.2 | Enrichment of $MT_\Delta$ by $\overline{A}$ .......................... 56 |
|  |  | 3.3.3 | Fundamental Domain of a $PT$ ...................... 60 |

**4  Species of Structures and Axiomatic Basis of a $PT$** ................................................................. 63
  4.1  Mathematical Structures ................................. 64
  4.2  Deduction of Structures ................................... 67
  4.3  Axiomatic Basis and Fairy Tales.......................... 73
  4.4  Pure Laws of Nature....................................... 76
  4.5  Change of the Mathematical Form of an Axiomatic Basis......................................... 78
  4.6  Inaccuracy Sets and Uniform Structures .................... 85
  4.7  Do the "Laws of Nature" Describe Realities?................ 92
  4.8  Classification of Laws of Nature ........................... 96
  4.9  Skeleton and Uninterpreted Theories ...................... 101

**5  Relations Between Various $PT$s** .......................... 105
  5.1  Relations Between Two $PT$s with the Same Application Domain ............................................... 106
  5.2  Relations Between Two $PT$s with a Common Part of an Application Domain ...................................... 111
  5.3  Pre-theories............................................... 112
  5.4  Relations Between $PT$s with Different Application Domains .. 116
  5.5  Approximation Theories................................... 117
  5.6  The Network of $PT$s ..................................... 118

**6  Real and Possible as Physical Concepts** .................... 121
  6.1  Closed Theories ........................................... 124
  6.2  Physical Systems .......................................... 128
  6.3  New Concepts in a $PT$ ................................... 131
  6.4  Indirect Measurements .................................... 133
  6.5  Classifications and Interpretations ......................... 138
  6.6  The Reality Domain of a $PT$ ............................. 143

---

**Part II Examples of Simple Theories**

---

**A  A Description of the Surface of the Earth, or of a Round Table** ......................................... 147

**B  A Simplified Example of Newton's Mechanics**.............. 159

C   **The Structure of the Human Species** ....................... 167

**References** .................................................... 173

**List of Symbols** ............................................... 175

**Index** ......................................................... 177

# Intention of the Book

## Motivation and Problem Setting

The aim of this book is to give a description of a method of formulating physical theories. The reason for the development of such an analysis is that many obvious ambiguities in physics have shown the necessity of understanding in a critical manner the different formulations of physics, in particular theoretical physics.

During the edification of a science, "preliminary decisions" intervene which are not shared by everyone. Consequently, it is desirable to formulate and to name as well as possible these preliminary decisions, not to reject other ideas of physics as meaningless or "reasoning errors," but to show that there is no construction of a science without preliminary decisions and that the method proposed here is *one* possibility for better understanding physics as a science (at least this is what the authors hope for).

Historically, a science does not begin its development by reflecting on its foundations. It starts rather with the accumulation and the assimilation of new knowledge. Its methods are intuitively conceived and applied in a fruitful way. But at a given time one meets with contradictions. These contradictions must be clarified if one intends to develop science in a serious way. This is carried out by seeking to discover the cause of these contradictions. Once identified, one then tries to specify the methods of the concerned science, so that these contradictions can be avoided. As long as these contradictions exist, the methods leading to these contradictions are, at least in the beginning, provided with warnings that make it possible to avoid them with a certain prudence.

The appearance of contradictions in physics is such a "common" characteristic of its development that we are almost not aware of the fact anymore. But it is precisely these contradictions that contribute to the development of physics, and the greater the contradictions, the greater the success after having overcome them. The numerous more or less important contradictions between the theory and the reality are always new impulses making it possible to improve the physical theories. Two major contradictions in physics were, e.g., the divergence between the concept of space–time initially used and the results of the Michelson's experiment, leading to the development of the special relativity theory (see [2, Chap. IX]), and the divergence between the classical model of the atom and the quantum emission of light (the contradiction between the corpuscular and undulatory theories), leading to the development of quantum mechanics.

The contradictions of set theory in mathematics, as well as the contradiction existing between the complementary terms "wave" and "corpuscle" in physics, continue to be the cause of philosophical quarrels about the nature of mathematics and physics, respectively. Although these questions are interesting and justified, the philosophical discussions contribute only very little to the improvement of the methods of the concerned sciences. On the other hand, each particular method of a science presupposes – generally in an unconscious way – certain philosophical conceptions that influence the structure of the method. In spite of this, it is not common to refer to philosophical arguments in order to justify the method; only the "success" of the methods of the concerned science is preponderant. Thus, all the philosophical doubts about the mathematical concept of infinity were not able to change anything in the fast development of the mathematics of infinite systems. All the questions regarding the objectivity of the world have not prevented experimental physicists from regarding their results of measurement as objective facts.

One cannot force someone to accept mathematics if, on the basis of presuppositions of a philosophical nature, he refuses these methods, e.g., the classical logic. It is also not possible to make someone accept the methods of physics, just as they are effectively used, if that person refuses a priori the possibility that one can observe objectively real facts, i.e., if that person does not want to accept, as a basis of science, *the completely normal and unconscious behavior of men with respect to their usual environment as a world of things objectively present and of events being held objectively.*

Our task will not be to philosophically justify the methods of physics, but to analyze them, to specify them, and to examine their structures. This does not mean that there should not be a philosophical reflection about physics and the methods of physics afterwards. *We will try to solve the first part of this problem by a formalization. The formalization abstracts the contents of significance in order to precisely fix each step in the "rules of the game." The second part will consist in examining the structure of this "game."* Concerning mathematics, the two aspects of the problem have already been largely dealt with. In this book, we wish to begin a similar essay for physics. In response to the objection that until now physics could exist without such an examination of its methods, one can refer to the example of the set theory in mathematics, where one first avoided the difficulty by using an intuitive concept of sets until a more precise analysis of the foundations had become necessary in order to eliminate the contradictions that appeared.

One cannot retain as an objection to the "rules of the game" of the physical methods given hereafter that until now these rules were not always complied with. But it is precisely in what the improvement consists in, that one can more precisely realize what will be allowed in physics.

We do not claim to formulate methods definitively, in such a way that contradictions will never appear again. We know today the problems of such "proofs of absence of contradiction" of a system. We only hope that the analysis to which we aspire will show that the methodical rules of the game are adapted to the "actual" problems of physics.

To summarize, we give names to each of the sets of problems outlined:

- *Formal methodology of physics* as a description of the "rules of the game" of physics
- *Fundamental physics* as an examination of the structure of the system of the methodical "rules of the game" of physics, and as an examination of the construction of physics as a whole or, better, of the various possibilities of construction of physics as a whole

## Current State of Art

Concerning the current state of art we refer to the work of Erhard Scheibe because of the many similarities between his theoretical conception of a physical theory and the "new form of physical theory" proposed in this book. In particular, we refer to his two books about the reduction in physics (see [3,4]), and also to the book *Between Rationalism and Empiricism* (see [5]), a representative selection of his writings on the philosophy of physics, in which other aspects of his theoretical conception are also covered.

## Main Ideas of Our Approach

The "new form of physical theory" proposed in this book is based on the following ideas:

### Reality and Facts

The reality is in part constituted of facts stating "basic properties" of objects and "basic relations" between objects. Only facts related to the "domains of physics" are taken into consideration. These facts, directly recordable or indirectly recordable via known theories, called *pre-theories*, constitute what we call the physically recordable domain or the *reality domain*. The directly and indirectly recordable facts are simply a collection of empirical evidences suggesting or confirming the existence of more interesting facts hidden behind them.

### Imagined Realities or Fairy Tales

We usually only observe a small fraction of the facts constituting the object of our investigation. Facts are like an iceberg; they are mostly submerged under the surface of immediate experience. The submerged part of these facts must be hypothesized. As long as a postulated system of hypotheses (asserted by means of propositions, implying new physical concepts) does not refer to the facts of reality, we shall speak of imagined realities or "fairy tales."

### Basic Language

Facts stating properties of objects and relations between objects are denoted by sentences formulated in a natural language of very simple form, called a *basic language*. The semantics of the basic language make it possible to clarify the relations between the linguistic, the conceptual, and the reality levels, i.e., between the sentences, the propositions, and the facts.

### Application Domain

The restriction of the reality domain to the facts (directly recordable or indirectly recordable via pre-theories) considered a priori is the *application domain* of the intended physical theory.

### Recording Process

Only facts related to the application domain are considered in the recording process. These facts are denoted under the form of a collection of sentences formulated in the basic language. In other words, only facts denoted by sentences using terms that designate property or relation concepts, belonging to the context related to the application domain, are taken into consideration.

### Mathematization Process

Natural sentences formulated in the basic language (related to the application domain) are transcribed into formal sentences formulated in a mathematical language. This formal language is that of the *standard mathematical theory*.

### Idealization Process

We know that practically all mathematical theories used in physical theories can only be approximations of the reality, i.e., they can be applied to an application domain only under the assumption of allowing for some degree of approximation or inaccuracy. The standard mathematical theory is then enriched by mathematical and physical idealizations, in order to obtain an *idealized mathematical theory*.

## Axiomatization Process

The idealized mathematical theory becomes especially significant through a "structuring" axiomatization of the theory. In this structuring axiomatization, the original objects and relations do not occur independently any more, but only as links in an overall structure, and the axiomatic system makes assertions about this overall structure. What is characterized by an axiomatic system is not a determinate structure, but a species of structures. At the same time, the same species of structures can, in general, be defined by means of several different axiomatic systems.

## Relations Between Physical Theories

Physics does not consist of only one theory; it is made up of a set of various theories. It is possible to establish relations between physical theories (with application domains that are either the same, partially the same, or completely different). A physical theory can be an approximation of another theory. It is also possible to build networks of physical theories.

## New Physical Concepts

The reality domain can be extended by hypotheses, i.e., by postulated relations between recordable and nonrecordable (or "imagined") facts. In terms of the semantic of the basic language, this is equivalent to (a) defining a new (class or relation) concept, (b) inventing a new word designating this new concept, and (c) imagining a new process set up with the purpose of obtaining a real reference to the new concept. In other words, to an extension that satisfies the semantic relations (of designation, reference, and denotation) corresponds an extension of the reality domain, in so far as the new conceptual entity (which has been hypothesized) refers to a new real entity. As long as the new concept has no real reference, we will speak of a "fairy tale concept" at the conceptual level, and simply of a "fairy tale" at the reality level.

# Outline of the Book

## Chapter 1. Reality

We describe what we call the "reality" related to a physical theory, the goal of which is to provide a satisfactory understanding of certain aspects of this reality. In particular, we state what we consider to be a real entity or only a "fairy tale."

## Chapter 2. Building of a Mathematical Theory

We outline the formal construction of a mathematical theory. It might appear superfluous to begin this book with a draft of a formal construction of a mathematical theory. The intention of this draft is not to give the reader a precise description of the various possibilities of a construction of mathematics, but simply to elucidate how the three fundamental parts, logic, set theory, and species of structure, *join together* for the construction of a mathematical theory.

## Chapter 3. From Reality to Mathematics

We are concerned with the problem of correspondence between the structure of a part of reality and an idealized mathematical structure that is similar to the structure of that part of reality. We will not explain or base this structure on philosophical or other points of view, but by the results of experiments. The only foundation of a physical theory is the success of the method of physics that we will describe here in detail. The formulation of this method is described by considering the transition from reality to mathematics, distinguishing three processes: (a) A recording process, which is a formulation of recorded facts denoted under the form of sentences in a natural language of very simple form called the basic language of the intended physical theory; (b) A mathematization process, which is a transcription of natural sentences formulated in the basic language into formal sentences formulated in a formal language (expressing the mathematical theory); (c) An idealization process, which is an enrichment by mathematical and physical idealizations of the previous mathematical theory.

## Chapter 4. Species of Structures and Axiomatic Basis of a *PT*

We are concerned with the axiomatization of the idealized mathematical theory. The best way in which to reveal the deep structures of the idealized mathematical theory is by means of formal analysis which lead to more precise conceptual entities. One of these formal tools appears to be the set theory. Our method of viewing the deep structures refers to the metatheoretical structuralism approach to structures, where the term "structure" is understood in the sense of Bourbaki (see [6, Chap. IV]). It should be clear that the set theory is only the form and not the substance of the theory. In principle, other methods of analysis could be used, e.g., the "category theory."

## Chapter 5. Relations Between Various *PT*s

Here we introduce the idea that physics consists not only of one theory, but it is made up of a set of various theories. Diverse modes of relation between physical theories are taken into consideration. We also introduce the notion of physical theories connected within a network.

## Chapter 6. Real and Possible as Physical Concepts

In this last chapter, we provide an answer to the question "How can we state real facts with the help of a physical theory even if they were not stated before by direct observations or with the help of pre-theories?"

Throughout Part I, the introduced concepts will be illustrated on the basis of a simple example, noted *Example A*: *A description of the surface of the earth, or of a round table*. Furthermore, other examples are given in Part II.

# Part I

# A New Form of Physical Theory

# 1
# Reality

In this chapter, we will be concerned with the reality related to a physical theory, the goal of which is to provide a satisfactory understanding of certain aspects of this reality.

## 1.1 The Structure of Reality

We assume that reality is in part constituted of facts stating "basic" *properties* of objects and *relations* between objects. Furthermore, we assume that the facts, directly or indirectly (via pre-theories, see Sect. 5.3) recordable, are of the following form:

- the object $\bar{a}$ has the property $\widetilde{p}$;
- between the objects $\bar{a}_1, \ldots, \bar{a}_n$ and finite many real numbers $\alpha_1, \ldots, \alpha_n$, there is the relation $\widetilde{r}_n(\bar{a}_1, \ldots, \bar{a}_n, \alpha_1, \ldots, \alpha_n)$.

Only facts related to the "domains of physics" are taken into consideration. These facts constitute what we call the *physically recordable domain*, or *reality domain*, and is denoted by $W$.

This reality domain $W$ can be extended by hypotheses. To the facts, directly and indirectly recordable, is "added" a process set up with the purpose of testing hypotheses by means of *experiments*, which have the purpose of providing (at least to a certain extent) a real reference to the postulated hypotheses, i.e., to provide a real reference to the postulated relations between the recordable and nonrecordable (or "imagined") facts.

The directly and indirectly recordable facts are simply a collection of empirical evidences suggesting or confirming the existence of more interesting facts behind them. Facts are like an iceberg; they are mostly submerged under the surface of immediate experience. The submerged part of these facts must be hypothesized. In order to test such hypotheses, relations between the recordable and nonrecordable facts must be added, by which the recordable facts can count as evidence for or against the existence of the nonrecordable facts.

## 1.2 The Physical Reality

In our conception of a physical theory, we consider three domains of *physical reality* corresponding to particular conceptual levels of a physical theory. In the case of a particular physical theory, denoted by $PT_\nu$, we distinguish the *application domain*, denoted by $A_{p_\nu}$, the *fundamental domain*, denoted by $G_\nu$, and the *reality domain*, denoted by $W_\nu$. Let us briefly describe these three domains.

### 1.2.1 The Application Domain of a *PT*

The application domain of a particular physical theory $PT_\nu$, denoted by $A_{p_\nu}$, is the restriction of the reality domain $W$ to the facts that the theory considers a priori.

The recording of facts can be made directly or indirectly by pre-theories. It is similar to the reading of a text. The application domain $A_{p_\nu}$ is limited by the domain of physical concepts that are to be used for that reading. The restriction of $A_{p_\nu}$ by the domain of concepts is essential since there is no physical theory for the whole of reality. Common (or contextual) domains of physical concepts have, e.g., the following designations: mechanics, optics, thermodynamics, electrodynamics, etc.

We shall see later (Sect. 3.1.2) that only facts denoted by sentences using terms that designate physical concepts, belonging to the context related to the application domain $A_{p_\nu}$, can be recorded.

From a methodological point of view, $A_{p_\nu}$ is something given a priori relative to the physical theory $PT_\nu$. Something which is given a priori is not implicitly defined, it can only be shown.

By this action of showing, one does not wish to consider only directly recordable facts, denoted by $\rho$, but also indirectly recordable facts stated by other physical theories $PT_\alpha, PT_\beta, \ldots$ (but of course *not* by the physical theory $PT_\nu$ under examination). For example, an electrical current in a conductor can be considered as a fact for a $PT_\nu$, i.e., can be part of the application domain $A_{p_\nu}$, even though it is only by electrodynamics that one can speak of currents as given facts. The physical theory $PT_\nu$ in consideration, in which such a current belongs to the application domain $A_{p_\nu}$, *cannot* naturally be electrodynamic, but a $PT_\nu$ which presupposes electrodynamics in the way mentioned above (e.g., quantum mechanics). On the other hand, if one wishes to regard electrodynamics as a $PT_\nu$, a current does not belong to the application domain $A_{p_\nu}$. Other "demonstrable" facts such as forces (defined by mechanics) belong in this case to the application domain $A_{p_\nu}$. It is only by a $PT_\nu$ (i.e., by the electrodynamics) that a current in a conductor becomes part of the reality domain $W$ (see, e.g., [2] VIII).

The basis of all observations is the possibility to admit certain facts in an immediate experience, i.e., without any $PT_\nu$. For example, in an experiment, the state of a counter is accepted as a fact and does not need to be analyzed. One uses no scientific criterion, or even physics, to regard such facts as being certain.

It is decisive that the question of the justification of regarding such given facts in an immediate experience as real facts is *neither posed nor solved* by physics; it is by the *exclusion of this question* that physics as such is possible. The nonphysical question of the recognition of such given facts is not a question of criterion, but a complicated question of physical, physiological, psychological, and cognitive processes that we cannot entirely realize, such as, e.g., the case that this afternoon at the time of our walk a hare crossed our path. We can base our certainty neither on the depositions of others (who were not present at the time) nor on photographs (which were not taken) nor on some other "criteria." Naturally, physics does not formulate any objection with regard to the treatment of this question of knowledge of the given facts in a pre-physical domain.

However, there arises a very interesting and significant question in fundamental physics: Is the starting point of the given facts for all of physics consistent with the physics which is developed by it? This problem of consistency between the physics developed on the basis of given facts and the "physical representation" that results from the processes of sensory perception is "described" in more detail in [2, Chap. XVII]. There has existed, at least until now, no indication that such a consistency between physics and sensory perceptions is not given.

To summarize, in the definition of the application domain $A_{p_\nu}$ of a physical theory $PT_\nu$, one can already include the reality domains $W_\alpha, W_\beta, \ldots$ of other physical theories $PT_\alpha, PT_\beta, \ldots$. We call these $PT_\alpha, PT_\beta, \ldots$ *pre-theories* of $PT_\nu$. The definition of the application domain $A_{p_\nu}$ is thus not trivial. It is a problem of *fundamental physics*, which we will only be able to approach later. This application domain $A_p$ will be further explained in Sect. 3.1.2.

### 1.2.2 The Fundamental Domain of a *PT*

The fundamental domain of a particular physical theory $PT_\nu$, denoted by $G_\nu$, is the restriction of the application domain $A_{p_\nu}$ to the facts that the theory describes.

We know that practically all mathematical theories used in physical theories can only be approximations of the reality, i.e., they can be applied to an application domain $A_{p_\nu}$ only under the assumption of allowing for some *degree of approximation* or *degree of inaccuracy* (see Sect. 3.3).

The fact of having a usable theory depends on the choice of the degree of inaccuracy allowed. It is necessary to distinguish between two cases:

- We have no large inaccuracies, and we say that the theory can be applied as a "good" description of the application domain $A_{p_\nu}$;
- We have large inaccuracies, and we say that the theory says practically nothing about the structure of the reality in such regions.

But why will we then apply this theory onto the total application domain $A_{p_\nu}$? It is often more useful to apply a theory only on that part of the application domain $A_{p_\nu}$ where we can use a small degree of inaccuracy. In such a region the theory essentially says something about the structure of reality and will be useful for technical applications. We call such a region (a part of the application domain $A_{p_\nu}$) the *fundamental domain* $G_\nu$. If we can use "small" inaccuracies in the total application domain $A_{p_\nu}$, then $G_\nu = A_{p_\nu}$. This fundamental domain $G_\nu$ will be further explained in Sect. 3.3.3.

### 1.2.3 The Reality Domain of a *PT*

The reality domain of a particular physical theory $PT_\nu$, denoted by $W_\nu$, is the extension of the fundamental domain $G_\nu$ to the facts (related to the new physical concepts) that the theory describes.

Our task is not only to detect "nonmeasured" realities, but also to detect new realities.

We have to introduce new words designating new physical concepts in order to denote these possible or imagined realities. But we do not say how we can "observe" such possible realities.

How to introduce new physical concepts will be further explained in Sect. 6.3.

To observe "basic" *properties* of objects and *relations* between objects, we can use immediate observations and "pre-theories," i.e., only well-defined methods, before the introduction of a mathematical theory. This problem is much more difficult than the introduction of new physical concepts. If we want to determine the factual reference of the new physical concepts, then we must go back from the extended mathematical theory to a reality domain $W_\nu \supset A_{p_\nu}$. This reality domain $W_\nu$ will be further explained in Sect. 6.6.

### 1.2.4 The Reality Domain of all *PT*s

As we have seen before, only facts belonging to physical domains are taken into consideration and constitute what we call the *physically recordable domain*, or *reality domain*, and is denoted by $W$.

We can now add that the *reality domain* $W$ is the domain of all $W$s, i.e., the $W$s of all *PT*s. Given that all *PT*s are not known, $W$ cannot be established. By finding new *PT*s, one discovers new $W$s (e.g., atoms and elementary particles). The *physically recordable domain* $W$ remains decisively limited by the fact that one does not permit all directly ascertainable facts such as, e.g., that a sound is harmonious or that a violin has a good sound.

The domain of attainable facts certified for physics has not been established until this day. This reality domain $W$ will be further explained in Sect. 6.6.

### 1.2.5 Remarks

The methods of establishing $PT$s are also applicable to completely different areas than physics, for example, our theory about the structure of the human species (see Part II, Example C) or, more interestingly, the application of Weidlich's theory to sociological problems (see [7]).

Figure 1.1 represents a summary of the domains of physical reality.

$W$
**reality domain of all $PT$s**
*all*
*directly and indirectly (via pre-theories) physically recordable facts*

$\cup$

$\underbrace{\rho, W_\alpha, W_\beta, \ldots}_{A_{p_\nu}}$
**application domain of $PT_\nu$**
*restriction of $W$*
*to the facts that the $PT_\nu$ a priori considers*

$\cup$

$G_\nu$
**fundamental domain of $PT_\nu$**
*restriction of $A_{p_\nu}$*
*to the facts that the $PT_\nu$ describes*

$\cap$

$W_\nu$
**reality domain of $PT_\nu$**
*extension of $G_\nu$*
*to the facts, related to new physical concepts, that the $PT_\nu$ describes*

$\cap$

$W'$
$\overbrace{W, W_\nu}$
**reality domain of all $PT$s**
*all*
*directly or indirectly (via pre-theories) physically recordable facts*

**Fig. 1.1.** Domains of physical reality

## 1.3 Fairy Tales

Our task is not only to detect nonmeasured physical realities, but also to detect new physical realities.

For a particular physical theory $PT_\nu$, we have introduced the reality domain $W_\nu$ as the extension of $G_\nu$ to the facts related to the new physical concepts that the physical theory describes.

As long as a postulated system of hypotheses, or a postulated theory (expressed by means of propositions implying the new physical concepts), does not refer to the facts of reality, we shall speak of a *fairy tale theory* "at the conceptual level" and of an *imagined reality* or *fairy tale* "at the reality level." In Chap. 6 we will see that in general, one must consider a fairy tale theory as only physically possible.

There are many such fairy tales, or myths, in quantum mechanics. An example of such a fairy tale, which has not been established as being real (at least until now), is the very widespread idea that each microsystem has a real state that can be represented by a vector in a Hilbert space, e.g., a Schrödinger wave function.

And yet, to start from the idea of a fairy tale proves to be a very useful way in which to guess a physical theory, even if one runs the risk of introducing prejudices into such theories. One has often tried to prescribe principles to which the imagined facts should suffice, and sought to base these on philosophical considerations.

# 2
# Building of a Mathematical Theory

In addition to the physical reality (see Sect. 1.2), the mathematical theory, denoted by $MT$, is the second most significant part of a physical theory. As briefly as possible, we will describe the elements necessary for the construction of a mathematical theory. For a more detailed description, we refer, e.g., to [6].

"Mathematics deals with imagined objects and imagined relations between these objects." In order to clarify this assertion, one tries to formalize the methods and the results of mathematics, i.e., one formally establishes the structure of a mathematical text in order to clearly indicate what one understands by "terms," "relations," "axioms," "proofs," "theorems," etc. Since the unfolding of these formal methods is very close to what our intuition suggests, we will restrict ourselves in order to be able to give a concrete meaning to concepts such as "structure," "partial structure," "relation," etc. Without this mathematical construction, assertions such as "In a $PT$, a partial structure of an $MT$ gives us a picture of a real structure of the reality" would only have a very vague significance.

It is for this reason that we now take the time to describe the formal construction of an $MT$. Since in mathematics this cannot be done in an entirely homogeneous manner, it is still even less possible to give an overall picture of the various possibilities of such a formal construction. We have chosen a possibility that seems the most appropriate to our ends [6], namely the application of an $MT$ in a $PT$ (see Chap. 3).

## 2.1 Formal Language

A mathematical theory $MT$ is defined as an assembly of signs comprising certain rules. That one can define an $MT$ results from the fact that the formulation of all mathematical expressions, i.e., the mathematical language, is possible with a few very simple rules. Thus, conversely, in a formal manner, one can define an $MT$ by these rules of language, i.e., the rules for the signs.

One could call these rules, using a linguistical term, the *syntax* of the mathematical language. Our first task will be thus to describe this "syntax." We will describe it as the *rules of the game* with *signs*.

The signs that form the mathematical text are on the one hand letters, and on the other hand, recognizable signs such as, e.g., $\vee, \neg, \in, \subset$. The signs are joined together under the form of assemblies of signs, where an assembly is a succession of signs written next to one another. First of all, one introduces rules that must characterize the "well-formed" assemblies according to the syntax of the mathematical language, and which make it possible, among these, to distinguish the assemblies that represent "objects" and the assemblies that represent "relations."

This manner of considering a mathematical text is particularly significant to physics. On the one hand, one wants to denote the facts of the application domain $A_p$ into "well-formed relations" formulated in the mathematical language of $MT$ (see Chap. 3). On the other hand, one also wants the relations of $MT$ to lead to assertions about the reality domain $W$ (see Chap. 6). Consequently, it is clearly necessary to lay down the rules according to which well-formed relations must be formulated in an $MT$.

To this end, we divide the signs into three categories:

1. The logical signs: $\vee, \neg, \tau$. The sign $\vee$ means "or," the sign $\neg$ means "not," and the sign $\tau$ means "an object which …." The sense of the sign $\tau$ will be explained more precisely hereafter. These logical signs are sufficient. Given that a precise development of all the rules is not significant for us, we will write from now on more concretely: instead of $\neg A$ always "not $A$;" instead of $\vee AB$ always "$A$ or $B$;" instead of $\vee \neg AB$, i.e., instead of "(not $A$) or $B$" always "$A \Rightarrow B$," or in words "$A$ implies $B$;" and for "not [(not $A$) or (not $B$)]" simply "$A$ and $B$."
2. The letters; they always represent objects. Assemblies can also represent objects.
3. The specific signs of the $MT$ considered as, e.g., the sign $\in$ of the set theory.

Only the assemblies that result from the following rules are allowed in an $MT$ and are concretely accepted as well-formed assemblies. A characteristic must be attributed to each specific sign belonging to the third category. This is a sign, either substantific or relational, i.e., a sign that determines a term (concretely, an object) or a relation (concretely, an assertion about object(s)). Each one of these specific signs must still receive a weight, an integer $n$.

One designates by "terms" of the $MT$ all assemblies that begin with a $\tau$ or a substantific sign, or which consist only of one letter, and one designates by "relations" of the $MT$ all the other assemblies.

One designates by "formative construction" in an $MT$ a sequence of assemblies that have the following property, i.e., for each assembly $A$ of the sequence, one of the following conditions is satisfied:

(a) $A$ is a letter.
(b) $A$ is identical to "not $B$," $B$ being a relation preceding $A$ in the sequence.
(c) $A$ is identical to "$B$ or $C$," $B$ and $C$ being relations preceding $A$ in the sequence.
(d) $A$ is identical to $\tau_x(B)$, $B$ being a relation preceding $A$ in the sequence and containing the letter $x$. To indicate this, we often write $B(x)$ instead of $B$. One can concretely interpret this by "$x$ is an object in an assertion $B(x)$" (e.g., $x \in M$, i.e., $x$ is an element of the set $M$). The term $\tau_x(B)$ is a "privileged" term which, inserted in $B(x)$, satisfies the relation $B$ (e.g., $\tau_x(x \in M)$ is a privileged element of the set $M$).
(e) $A$ is identical to $sA_1 \cdots A_n$, $s$ being a specific sign of the third category, of weight $n$, and $A_1$ to $A_n$ being terms preceding $A$ in the sequence. If $s$ is a substantific sign, then $sA_1 \cdots A_n$ is a new object constituted of objects (an assertion over the objects) $A_1 \cdots A_n$.

## 2.2 Axioms and Proofs

The rules described up to now are only used to characterize the well-formed assemblies. Now we must indicate the methods according to which one decides if an assertion (to speak concretely) is "true." This is done by the posing of axioms and by the building of proofs. In mathematics the axioms are, so to speak, true assertions by definition. If these axioms become assertions on the reality domain $W$ (see Chap. 6), then the truth of an axiom in $PT$ takes a new sense with respect to $MT$. For this reason, we do not want to speak about "true" and "false" in an $MT$, as is often the case, since the axioms are posed and are not the result of an act of knowledge. The "truth" of many axioms cannot be perceived, since such a truth does often not exist because in an $MT$ it is possible to pose, instead of the axiom $A$, the axiom "not $A$" (as a "physical" example, see the "axiom of simultaneity," in [2, Chap. VII], and its "nonvalidity" in the theory of special relativity, in [2, Chap. IX]).

The posing of axioms is a process of decisive importance for mathematics, as well as for physics, which requires a general explanation. We distinguish the explicit axioms and the axiomatic rules below.

An *explicit axiom* is a relation written according to the rules of Sect. 2.1. More than one such axiom can be written. In these explicit axioms, some letters (concretely, indefinite basic objects of the $MT$) can appear. They are called constants of the $MT$. These explicit axioms concretely represent true assertions about these basic objects. But one can also say that the basic objects are implicitly defined by the axioms. In brief, one often gives names to these basic objects (as an abbreviation for the totality of the axioms posed for them).

Thus, e.g., a term $x$ (where $x$ is a basic object) denotes an "ordered set" if an ordering relation is defined on this term with corresponding axioms.

The *axiomatic rules* are not relations in the sense of Sect. 2.1, but rules from which one can obtain new relations starting from relations already present in the text. Intuitively it must provide "identically true" relations, i.e., whatever the relations used in the application of an axiomatic rule, one obtains (intuitively) a "true" relation. We will meet such axiomatic rules, e.g., in Sect. 2.3, as logical rules (i.e., intuitively as a combination of logically identical true relations).

The simplest way in which to express the axiomatic rules is to use abbreviations for the assemblies. An axiomatic rule can be written as a symbolic relation constituted of these "abbreviations." These symbolic relations are also called *implicit axioms*. Letters used as abbreviations do not really appear in the theory since "any" relations resulting from the theory can be used in their place. A mathematical theory $MT$ then consists of a text of distinct relations (concretely, "true" or "valid" assertions) that can be obtained using the following three rules:

1. the explicit axioms themselves;
2. the implicit axioms, if they contain terms and relations built according to the rules of Sect. 2.1;
3. of a relation $B$, in the case where the two relations $A$ and "$A \Rightarrow B$" appear previously in the text of the $MT$.

All of the relations resulting from (1) to (3) (concretely, "true" assertions compared to the only well-formed assertions in the sense of Sect. 2.1) are called *theorems* of $MT$. For greater practicality, we include all the explicit axioms to the theorems of $MT$.

If a well-formed relation (i.e., formed in the sense of Sect. 2.1) cannot be obtained with the three preceding rules, then it is *not a theorem* in $MT$. Let us note that the fact that the relation $A$ is not a theorem in $MT$ *does not imply* that "not A" must be a theorem in $MT$. This will also be significant in the development of physical theories and, in particular, in the transition to more extended theories, and in the appreciation of the "physical reality" of facts not observed (Chap. 6).

One already introduces here another concept which will become very significant in a $PT$. This concerns the comparison of two $MT$s. A theory $MT_2$ is said to be "stronger" than a theory $MT_1$ if all of the signs of $MT_1$ are signs of $MT_2$, all of the explicit axioms of $MT_1$ are theorems in $MT_2$, and all of the implicit axioms of $MT_1$ are implicit axioms of $MT_2$. It follows that all the theorems of $MT_1$ are theorems in $MT_2$.

The transition from an $MT_1$ to a stronger $MT_2$ will become of paramount importance during the construction and the extension of a $PT$, because the stronger the $MT$ becomes, the more the $PT$ on which it depends becomes more expressive.

## 2.3 Logics

The first implicit axioms to be introduced are related to logic. Here one decides to use "normal," "bivalent" logic and not a polyvalent or other form of logic. Given that particular relations of an $MT$ become in a $PT$ assertions on the reality, we thus presuppose this logic for all $PT$, which will become clearer in Chaps. 3 and 6.

Attempts to modify logic were tried in mathematics as well as in physics. Quantum mechanics was used in physics as an argument for the need of a polyvalent logic, a probability logic with a continuous scale of values having "true" and "false" as limiting extremes. The fact that we build a quantum theory with normal logic shows that such a necessity does not exist.

We introduce logic by the following axiomatic rules: If $A$, $B$, $C$ are relations, then the relations

$$(A \text{ or } A) \Rightarrow A, \tag{2.3.1}$$

$$A \Rightarrow (A \text{ or } B), \tag{2.3.2}$$

$$(A \text{ or } B) \Rightarrow (B \text{ or } A), \tag{2.3.3}$$

$$(A \Rightarrow B) \Rightarrow ((C \text{ or } A) \Rightarrow (C \text{ or } B)) \tag{2.3.4}$$

are implicit axioms of $MT$.

If one considers a relation with *two* possible values, "true" or "false," and one attributes to the relation "$A$ or $B$" the value true if at least one of the two relations $A$ or $B$ is true, otherwise the value false, and one attributes to the relation "not $A$" the value true if $A$ is false, and conversely, then (2.3.1) to (2.3.4) represent true relations (because "$A \Rightarrow B$" means by definition "(not $A$) or $B$," i.e., "$A \Rightarrow B$" is true if $A$ and $B$ are true or $A$ is false).

And yet one should not confuse the logical implicit axioms (2.3.1) to (2.3.4) with the intuitive association of "true" or "false" to "any" relations $A, B, \ldots$. However, in Sect. 2.2 we had not introduced the "values" true and false for relations, but only laid down the rules of proof deriving new relations from axioms. What one would express intuitively is as follows: In this $MT$, "$A$ is a true relation" is replaced by the new expression "$A$ is a theorem in $MT$." We have already outlined above that if $A$ is not a theorem in $MT$, then it does not result inevitably that "not $A$" must be a theorem in $MT$. It may be thus that neither $A$ nor "not $A$" are theorems in $MT$. It is only in the following way that the logic introduced by the implicit axioms described above is a normal logic:

(a) If a relation $A$, as well as "not $A$," is a theorem in $MT$, then each well-formed relation $B$ (in the sense of Sect. 2.1) is a theorem in $MT$. Such an $MT$ is contradictory and is unusable, since it does not in fact state anything. We will see in Chap. 3 that such a contradictory $MT$ already leads to a *completely* "unusable" $PT$ before one has *tested such an MT by*

22    2 Building of a Mathematical Theory

*experiment.* For this reason, we eliminate all the contradictory $MT$. Then only $A$ or only "not $A$" can be a theorem in $MT$.
(b) The following principle of *proof by contradiction,* often used during proofs, is also valid: If one adds to $MT$, as an additional axiom, the relation "not $A$," then one obtains a theory $MT'$ *stronger* than $MT$ (in the sense of Sect. 2.2); and if $MT'$ is contradictory, then $A$ is a theorem in $MT$.
(c) If $A$ is a theorem in $MT$, then "$A$ or $B$" is a theorem in $MT$.
(d) If $B$ is a theorem in $MT$, then "$A$ or $B$" is a theorem in $MT$.
(e) If "not $A$" and "not $B$" are theorems in $MT$, then "not $(A$ or $B)$" is a theorem in $MT$; and also if "$A$ or $B$" is a theorem in $MT$, then "not [(not $A$) and (not $B$)]" is a theorem in $MT$, i.e., "not $A$" as well as "not $B$" cannot be theorems in $MT$.

The two criteria (a) and (b) fix the sense of "not," and what we briefly indicate in the *new form* by bivalent logic.

The criteria (c) to (e) fix the sense of "or," which replaces the sense of "or" introduced above in an intuitive way, using the values "true" and "false." In short, we say that (c) to (e) fix the "normal" sense of "or."

In this sense, on the basis of (a) to (e), we say finally that the bivalent logic is introduced by the implicit axioms (2.3.1) to (2.3.4).

Of course, we will not draw here all the significant consequences for the technique of mathematical proofs of the implicit axioms (2.3.1) to (2.3.4) mentioned above (in particular to prove the deductions (a) to (e)). This is not necessary because the deductions obtained are for the most part "intuitively" obvious, and the reader is certainly accustomed to applying such logic and the methods of proof in mathematics. For a detailed description, we refer to [6]. In particular, it is easy to establish the above deductions (a) to (e) using the deductions mentioned in [6, Chap. I, Sect. 3.]. (a) is proved in [6, Chap. I, Sect. 3.1], (b) is identical to C 15 of [6, Chap. I, Sect. 3.3], and (c) and (d) easily result from the implicit axioms (2.3.2) and (2.3.3) mentioned above and the rule of proof (3) of the preceding section of this book (i.e., Sect. 2.2). (e) is a consequence of the equivalent relations mentioned under C 24 of [6, Chap. I, Sect. 3.5], "not (not $A$) $\Leftrightarrow A$" and "$(A$ or $B) \Leftrightarrow$ not [(not $A$) and (not $B$)]."

Because of their interest, particularly for the very significant reflection from the point of view of physics (see Chap. 3 and especially Chap. 6), two other relations which result from the implicit axioms (2.3.1) to (2.3.4) are added:

(f) Let $A$ be a relation in $MT$, and let $MT'$ be the theory obtained by adjoining $A$ to the axioms of $MT$. If $B$ is a theorem in $MT'$, then "$A \Rightarrow B$" is a theorem in $MT$. (*Proof:* see C 14 of [6, Chap. I, Sect. 3.3]).
(g) Let $A(x)$ and $B$ be relations in $MT$ ($x$ is not a constant of $MT$), let $T$ be a term such that $A(T)$ is a theorem in $MT$, and let $MT'$ be the theory obtained by adjoining $A(x)$ to the axioms of $MT$ ($x$ is thus a constant of $MT'$). If $B$ is a theorem in $MT'$, then $B$ is a theorem in $MT$. (*Proof:* see C 19 of [6, Chap. I, Sect. 3.3]).

From the two relations (a) and (b) above, for an $MT$ without contradiction, it results that if neither $A$ nor "not $A$" are theorems in $MT$, then one can add to $MT$ the relations $A$ as well as "not $A$" as axioms, and in this manner thus obtain two theories, $MT_1$ and respectively $MT_2$, which are both stronger than $MT$ and without contradiction. As already mentioned above, this situation is very significant for physics. In particular, we return to the discussion of the relation between the Galileo–Newton space-time theory and the special relativity theory in [2, Chap. IX].

We will not examine here the problems of the "proof" of noncontradiction in an $MT$. We adopt the point of view that the $MT$s used are without contradiction as long as a contradiction is not discovered. In the case where there would be a contradiction in $MT$, we would have to change the axioms in order to eliminate it.

If $A$ and $B$ are relations, we briefly write for the relation "$(A \Rightarrow B)$ and $(B \Rightarrow A)$":"$A \Leftrightarrow B$" and we say that "$A$ is *equivalent* to $B$." For any relations, because of the axioms introduced above, the following equivalences (C) are valid (as theorems in $MT$, see [6, Chap. I, Sect. 3.3]):

$(A \text{ and } (B \text{ or } C)) \Leftrightarrow ((A \text{ and } B) \text{ or } (A \text{ and } C))$,

$(A \text{ or } (B \text{ and } C)) \Leftrightarrow ((A \text{ or } B) \text{ and } (A \text{ or } C))$,

$(\text{not } (A \text{ and } B)) \Leftrightarrow ((\text{not } A) \text{ or } (\text{not } B))$, (C)

$(\text{not } (A \text{ or } B)) \Leftrightarrow ((\text{not } A) \text{ and } (\text{not } B))$,

$(\text{not } (\text{not } A)) \Leftrightarrow A$.

If we "formally" regard the sign $\Leftrightarrow$ as a sign of equality, and if we put the sign "$\wedge$" instead of "and" and the sign "$\vee$" instead of "or," then the logical theorems written above enter *formally* into the rules of calculation for a complemented distributive lattice. The fact that

$$(A \Rightarrow B) \Leftrightarrow [(A \text{ or } B) \Rightarrow B]$$

is a theorem can formally be interpreted so that the sign $\Rightarrow$ is the order determined by the lattice operations $\wedge$, $\vee$, so that also conversely the sign $\Leftrightarrow$ becomes formally a sign of equality.

However, there is a completely *decisive* difference between a complemented distributive lattice and the logical theorems above. The letters that appear in the logical relations are not elements of a set, and can in no way be applied directly to the logical relationship between relations. Relations in $MT$ and elements of a set are essentially different objects and should not be confused.

In spite of this, the five relations above (C) can be considered as the most concrete formulation used to introduce the classical logic in $MT$ by the implicit axioms (2.3.1) to (2.3.4). In what follows, we will always presuppose the validity of the implicit axioms (2.3.1) to (2.3.4).

Having introduced the logical sense of "or" and "not" in $MT$, one must emphasize that nothing has been said about the manner of logically binding relations related to facts of reality, because some such relations are first of all not combinations of signs appearing as relations $A, B, \ldots$ in $MT$ (see Sect. 2.1). It is only in Chap. 6 that we will speak of the problem of a relation between an $MT$ and the reality domain $W$, and by this also of the problem of the interpretation of the signs $\vee$ and $\neg$ of Sect. 2.1. With regard to the logical signs, the mathematization process (cor) is not simple; this is seen particularly with the sign $\tau$ introduced in Sect. 2.1, which will be further outlined in this chapter.

Whereas $\vee$ and $\neg$ obtained their sense by the implicit axioms (2.3.1) to (2.3.4) (concretely as "or" and "not"), we must now give a sense to $\tau$ by axiomatic rules. Previously, we introduced some abbreviations which have an obvious concrete sense. If $R$ is an assembly containing the letter $x$, then one can form the assembly $\tau_x(R)$ which does not contain any $x$ (see Sect. 2.1 and [6, Chap. I, Sect. 1.1]). If one substitutes $x$ by the assembly $\tau_x(R)$ in $R$ (i.e., everywhere where $x$ appears in $R$), then one obtains a new assembly which we denote by $(\exists x)R$. The assembly $(\exists x)R$ does not contain any $x$. The assembly $\tau_x(R)$ is concretely an object which satisfies $R$. The assembly $(\exists x)R$ is thus $R$ with a "particular object which satisfies $R$" put in the place of $x$. We can also say that "there exists an object which satisfies $R$." If $R$ is a relation, then $(\exists x)R$ is a relation ($\tau_x(R)$ is a term), i.e., according to Sect. 2.1, $(\exists x)R$ can appear in $MT$ only if $R$ is a relation. The fact that "there is not an object which satisfies (not $R$)" is concretely expressed by "$R$ is valid for all the objects." For this reason we shorten "not $((\exists x)(\text{not } R))$" by $(\forall x)R$. If $R$ is a relation, then $(\forall x)R$ is a relation, and it is meaningful in $MT$.

We now introduce the meaning, corresponding to the intuitive sense, of $(\exists x)R$ by an axiomatic rule.

If $R(x)$ is a relation containing the letter $x$ and if $T$ is a term, then

$$R(T) \Rightarrow (\exists x)R(x) \tag{2.3.5}$$

is an implicit axiom. $R(T)$ is the relation which results from $R(x)$ if $x$ is replaced by $T$. The implicit axiom (2.3.5) thus expresses the fact that there exists an object which satisfies $R$ if there is a $T$ which satisfies $R$.

For the details and consequences of the implicit axioms (2.3.1) to (2.3.5) introduced until now, one can refer to [6, Chap. I, Sect. 4], or one can follow the method exerted intuitively by using the expressions "there exists" and "for all."

However, some theorems will be mentioned without proof, since they will play a role in Chap. 3.

($\alpha$) If $R(x)$ is a theorem in $MT$ and if $x$ is not a constant of $MT$, then $(\forall x)R$ is a theorem in $MT$.

($\beta$) If $A(x)$ and $R(x)$ are relations of $MT$ ($x$ is not a constant of $MT$), and if $A(x) \Rightarrow R(x)$ is a theorem in $MT$, then "$(\exists x)A(x) \Rightarrow (\exists x)R(x)$" is a theorem in $MT$.

($\gamma$) If $A(x)$ and $R(x)$ are relations of $MT$, then the relations "$(\exists x)$ $(A(x)$ and not $R(x))$" and "$(\forall x)(A(x) \Rightarrow R(x))$" are equivalent.

By introducing the sign $\tau$, the relation $(\exists x)R$, the relation $(\forall x)R$, and the implicit axiom (2.3.5), we were not so concerned with the establishment of such theorems as such, but only with showing where these relations intervene in $MT$; however, one does not speak of physics or facts of reality. For this reason we warn, expressly and with insistence, not to identify blindly the expressions "there exists" and "for all" with "any forms of everyday assertion" on reality. To be able to express warnings here, we briefly give some examples of "common factual assertions" to which one will not give a sense during the construction of a $PT$ (at least not a directly obvious sense). It must be outlined that such factual assertions are not used as a basis for the construction of a $PT$ as is presented in this book.

Such meaningless assertions (at least for the moment) are, e.g., expressions such as "all ravens are black," "all electrons have the same mass $m$," "all men are mortal," etc. Let us take as an example the first expression: "all ravens are black." One can easily formalize it into the mathematical shape of Sect. 2.1: $r$ is a relational sign of weight 1 with the sense "to be raven," $s$ is a relational sign of weight 1 with the sense "to be black." The expression "all ravens are black" would then be written

$$(\forall x)(r(x) \Rightarrow s(x)).$$

But contrary to the everyday expression above, the relational signs $r$ and $s$ do not have significance, as for the contents (which is the intention) one cannot introduce the sign $\forall$ only "formally" as above in $MT$. Should "all ravens" also have a factual reference (and not only a formal one)? But what kind of a reference? What do we mean by "all ravens"? and how can "all ravens" be shown? In fact, where can "all ravens" be found?

We will never use in the construction of $PT$ expressions such as those mentioned above, except in *forms of shortened expressions*, but on which the logical rules are *not* applicable!

After this *warning*, "to not apply blindly forms of logical assertions, and logical rules of $MT$ to reality," we continue with the logical construction of an $MT$ by introducing the relation "to be identical" by a sign in $MT$, and by giving it a meaning in $MT$ using axiomatic rules.

Let us introduce as additional sign (for all the $MT$s which will be used later), the equality sign "$=$," a relational sign of weight 2 with the prescription, in the sense of Sect. 2.1, that "$= AB$" is a relation between two terms (i.e., two objects) $A,B$. Instead of "$= AB$" we write "$A = B$." For "not $(A = B)$"

we write "$A \neq B$." We fix the sense of the sign "=" by the following axiomatic rules:

If $R(x)$ is a relation, and if $A$ and $B$ are terms, then

$$(A = B) \Rightarrow (R(A) \Leftrightarrow R(B)) \tag{2.3.6}$$

is an implicit axiom.

If $R(x)$ and $S(x)$ are relations, then

$$[(\forall x)(R(x) \Leftrightarrow S(x))] \Rightarrow [\tau_x(R) - \tau_x(S)] \tag{2.3.7}$$

is an implicit axiom.

The implicit axiom (2.3.6) expresses the fact that it is "equal" if, in a relation $R(x)$ which includes the letter $x$, one replaces $x$ by $A$ or a $B$ identical to $A$, i.e., that the relations $R(A)$ and $R(B)$ are "the same" or – more precisely – they are equivalent. One can also say: Two identical terms $A$, $B$ also have the same "property" $R$. The implicit axiom (2.3.7) is not perceived in so intuitive a way, since the sign $\tau$ (concretely, "an object which...") is less easy to grasp in its intuitive content. The implicit axiom (2.3.7) says that for any $x$, two identical properties $R$ and $S$ imply that the term (concretely, object) determined by $\tau$ is identical for $R$ as well as for $S$, i.e., that the process of selection $\tau$ chooses "in the same manner" the identical properties $R$ and $S$.

For a detailed description of the consequences of this axiom we refer to [6] I, Sect. 5. Two theorems will be given, without proof, since they will often be used later and they will also have an importance from the physical point of view in Chap. 3. Let us begin with a definition:

If the relation

$$(\forall y)(\forall x)((R(y) \text{ and } R(x)) \Rightarrow (x = y))$$

is a theorem in $MT$ (it is often said that there exists at most one $x$ such that $R$), then $R(x)$ is said to be "single-valued in $x$" in $MT$. For each $MT$ which satisfies the axioms (2.3.1) to (2.3.7) there is the relation

($\delta$) If $R(x)$ is single-valued in $x$ in $MT$,

$$R(x) \Rightarrow (x = \tau_x(R))$$

is a theorem in $MT$.

And so, conversely, for a term $T$, the relation

$$R(x) \Rightarrow (x = T)$$

is a theorem in $MT$, then $R$ is single-valued in $x$ in $MT$.

Let us introduce another definition: If $R(x)$ is single-valued in $x$ and if

$$(\exists x) R(x)$$

is a theorem in $MT$, one says "there exists one and only one $x$ such that $R(x)$," and that $R(x)$ is "functional" in $x$ in $MT$. One has then

($\varepsilon$) If $R(x)$ is functional in $x$ in $MT$,

$$R(x) \Leftrightarrow (x = \tau_x(R))$$

is a theorem in $MT$.

And if, conversely, for a term $T$,

$$R(x) \Leftrightarrow (x = T)$$

is a theorem in $MT$, then $R(x)$ is functional in $x$ in $MT$.

In conclusion, we repeat once again that the fundamental logical axioms of an $MT$ (as summarized in Sect. 2.3) were not described to deduce as theorems of an $MT$ the known conclusions in mathematics, but were described to better distinguish later the sense of these axioms of an $MT$ in a $PT$ (e.g., in Chaps. 3 and 6).

## 2.4 Set Theory

Given that we presuppose also the set theory, we briefly turn our attention to the problem of the pose of axioms. We want especially to indicate some elements which will be significant for the use of these axioms in a $PT$. The problems of the use of the set theory in the representation $MT$ of a $PT$ can only be dealt with later; for this reason, almost no indication on the physical meaning will be given here for the moment.

In the set theory there appears as a new relational sign: "$z \in y$" (concretely, "$z$ is an element of $y$"). As an abbreviation for "$(\forall z)((z \in x) \Rightarrow (z \in y))$," i.e., for the relation "all the elements $z$ of $x$ are also elements of $y$," we briefly write "$x \subset y$" (concretely, "$x$ is part of $y$;" or "$y$ contains $x$" or similar expressions). For "not $(z \in y)$" respectively "not $(x \subset y)$," we often write "$z \notin y$" respectively "$x \not\subset y$."

The relational sign $\in$ will become of decisive importance for the use of an $MT$ in a $PT$. Concretely, a $PT$ includes assertions about the facts of reality as elements of a set (see Chap. 3).

For the set theory, it is decisive that the set is intuitively seen as the collective whole of all its elements. But, it is precisely this "whole of all" which is doubtful in physics, as we have already outlined in the introduction of the logical sign $\forall$ in $MT$. For example, the statement "set of all electrons" is not regarded as meaningful during the construction of a $PT$, because it is doubtful that this totality of all electrons exists.

In mathematics, "to collect in a set" is a significant notion of the set theory. If the set is a collection of its elements, then two sets must be identical if they have the same elements; for this reason, one requires as a first explicit axiom

$$(\forall x)(\forall y)((x \subset y \text{ and } y \subset x) \Rightarrow (x = y)). \tag{2.4.1}$$

But it is precisely this "collection in a set," intuitively so obvious, which leads to contradictions in mathematics when one neglects to take certain precautionary measures.

If we try to join together formally all the $x$ of a determined kind in a set, this can be carried out as follows: Let $R(x)$ be a relation, we shorten the relation "$(\exists y)(\forall x)((x \in y) \Leftrightarrow R(x))$" by "$\text{Coll}_x R$." If $\text{Coll}_x R$ is a theorem in $MT$ (the relation $R$ is said to be collectivizing in $x$ in $MT$), one says that the relation $R(x)$ determines a set. $y$ is the "set of all $x$ which satisfies $R(x)$," and because of "$(\forall x)((x \in y) \Leftrightarrow R)$" and "$(\forall x)((x \in z) \Leftrightarrow R)$," it follows the equality "$z = y$." For the relation "$S(y) \mid (\forall x)((x \in y) \Leftrightarrow R)$," there exists therefore at most one $y$ such that $S(y)$, i.e., according to Sect. 2.3, is single-valued in $y$. If $(\exists y)S(y)$ is a theorem in $MT$ (i.e. if $S(y)$ is functional in y), according to Sect. 2.3, then "$S(y) \Leftrightarrow (y = \tau_y(S))$" is also true. Consequently, if $\text{Coll}_x R$, i.e., $(\exists y)S(y)$, is a theorem in $MT$, we can denote the set of $y$ such that $S(y)$ by "$\tau_y[(\forall x)((x \in y) \Leftrightarrow R(x))]$," for that we write $\mathcal{E}_x(r)$ concretely "$\mathcal{E}_x(r)$ is the set of $x$ such that $R(x)$." The relation "$(\forall x)((x \in \mathcal{E}_x(R)) \Leftrightarrow R)$" is thus equivalent to $\text{Coll}_x R$, and the relation $R(x)$ is equivalent to "$x \in \mathcal{E}_x(R)$." Later on, the set $\mathcal{E}_x(r)$ will often be written in the usual form "$\{x \mid R(x)\}$."

But the set $\mathcal{E}_x(r)$ "exists" only if $\text{Coll}_x R$ is a theorem in $MT$. In no case, for all $R(x)$, is the relation $\text{Coll}_x R$ a theorem in $MT$. This seems curious, since there should "always" be, i.e., for each $R(x)$, the "set of $x$ such that $R(x)$." Wouldn't it be easy to conceive of $\text{Coll}_x R$ as an axiom for all $R(x)$? All those who have dealt with problems of an "intuitive" set theory know that such a general condition contains problems. For this reason we will proceed in a more careful way.

If one considers only the $x$ for which $R(x)$ is true and which are elements of a set $z$ ($z$ being susceptible to contain elements that do not satisfy $R$), then one expects that the $x$ which satisfy $R$ form a subset of $z$, i.e., $\text{Coll}_x R$ becomes a theorem in $MT$. If the relation $R$ still depends on an object $y$, and if all $x$ that satisfy the relation $R$ are elements of a set $z$ (which possibly depend on $y$), $y$ being fixed, then all $x$ that satisfy $R$, for at least an element $y$ of a set $u$, must form a set, which we require in the implicit axiom

## 2.4 Set Theory

$$((\forall y)(\exists z)(\forall x)(R \Rightarrow (x \in z)) \Rightarrow (\forall u)\text{Coll}_x((\exists y)((y \in u) \text{ and } R)). \quad (2.4.2)$$

This will make it possible to obtain, starting from sets, new sets using relations. But to be able to produce sets we pose the following axioms:

$$(\forall x)(\forall y)\text{Coll}_z(z = x \text{ or } z = y). \quad (2.4.3)$$

This means that if $x$ and $y$ are objects, then there is a set whose only elements are $x$ and $y$. We indicate this by $\{x, y\}$. This axiom is very easy to interpret in a *PT* as any finite set in which a finite number of $x_1 \cdots x_n$ are collected together. But for the *MT*, the infinite sets (which will be defined later) will be of great importance. These infinite sets made necessary a concrete axiomatization of the set theory; but it is not possible to interpret them physically, which we have already mentioned above, and which we will more precisely discuss in Sect. 3.2.4.

To continue to develop the set theory, we still need the possibility of introducing a pair $(x, y)$ of terms (objects) as a new term, i.e., a new object made up of two individual objects $x$, $y$

$$(\forall x)(\forall x')(\forall y)(\forall y')((x, y) = (x', y') \Rightarrow (x = x' \text{ and } y = y')). \quad (2.4.4)$$

The pair $(x, y)$ is different from the set $\{x, y\}$! In the pair, according to (2.3.4), the components $x$ and $y$ are ordered.

$$(\forall x)\text{Coll}_y(y \subset x) \quad (2.4.5)$$

means that "the set of all the subsets of a set $x$ exists." The last axiom

postulates the existence of an infinite set. (2.4.6)

An infinite set is precisely a set not finite. A finite set is defined by the fact that the cardinality changes if one adds one element to the set.

Each *MT* used in a *PT* is stronger than the set theory, i.e., all the axioms indicated until now are valid in *MT*. *In what follows, we suppose that each MT is stronger than the set theory.*

With regard to physics – as outlined in Sect. 3.2.4. – one could add to the set theory a seventh axiom that postulates that "there is no set whose size is strictly between that of the integers and that of the continuum." One could show that this seventh axiom is independent of the precedents. Otherwise, one could require that each set of an *MT* is either a set at most countable, or a subset of an echelon on sets at most countable.

In an *MT* (stronger than the set theory), starting from $n$ sets (terms) $E_1, \ldots, E_n$, one can build step by step new sets. We denote by $\mathcal{P}(E)$ the set of all subsets of $E$, and we denote by $E_1 \times E_2$ the set of all pairs $(x, y)$ with $x \in E_1$ and $y \in E_2$. When one applies a finite number of times the operations $\mathcal{P}$ and $\times$ to $E_1, \ldots, E_n$, one obtains new sets. Such a process, applicable in a finite number of steps, is called an *echelon construction* and the set thus obtained is called an *echelon*. The sets $E_1, \ldots, E_n$ are the base sets of the echelon construction. We denote an echelon by $S(E_1, \ldots, E_n)$, where the letter $S$ denotes the *echelon construction scheme*, whereby one obtained the echelon. If $E'_1, \ldots, E'_n$ are $n$ different sets, then $S(E'_1, \ldots, E'_n)$ is also an echelon of scheme $S$, but on the base sets $E'_1, \ldots, E'_n$.

In $MT$, let $f_i$ be mappings of the sets $E_i$ onto the sets $E'_i$, i.e., for any $x \in E_i$ one has $f_i(x) \in E'_i$, where $f_i(x)$ is defined for any $x \in E_i$. From the mappings $f_i$, one can then very easily build by *canonical extension* mappings of $E_1, \ldots, E_n$ onto $E'_1, \ldots, E'_n$. This is carried out step by step:

1. By defining a mapping $g$ of $\mathcal{P}(E)$ onto $\mathcal{P}(E')$ starting from a mapping $f$ of $E$ onto $E'$ such that, for a subset $e \subset E$, $g(e)$ is defined as the subset of all $f(x)$ such that $x \in e$.
2. By defining a mapping $g$ of $E_1 \times E_2$ onto $E'_1 \times E'_2$, starting from the mappings $f_1$ of $E_1$ onto $E'_1$ and $f_2$ of $E_2$ onto $E'_2$, by $g(x,y) = (f_1(x), f_2(y))$.

The application of $S(E_1, \ldots, E_n)$ onto $S(E'_1, \ldots, E'_n)$ thus obtained is denoted by $\langle f_1, \ldots, f_n \rangle^S$.

If all $f_i$ are injective (respectively surjective), then $\langle f_1, \ldots, f_n \rangle^S$ is also injective (respectively surjective), which one can easily show because this is valid for each step $\mathcal{P}$ or $\times$ of the echelon construction scheme $S$. If $f_i$ are mappings of $E_i$ onto $E'_i$ and $g_i$ of $E'_i$ onto $E''_i$, one denotes the mapping of $E_i$ onto $E''_i$ by $g_i f_i$. One has then

$$\langle g_1 f_1, \ldots, g_n f_n \rangle^S = \langle g_1, \ldots, g_n \rangle^S \langle f_1, \ldots, f_n \rangle^S.$$

If all $f_i$ are bijective (i.e., injective and surjective), with $g_i = f_i^{-1}$, then $\langle f_1, \ldots, f_n \rangle^S$ is also bijective and

$$(\langle f_1, \ldots, f_n \rangle^S)^{-1} = \langle f_1^{-1}, \ldots, f_n^{-1} \rangle^S,$$

where $f^{-1}$ is the inverse bijection of $f$.

If there are several elements $s_1, \ldots, s_p$ of any echelons $G_1, \ldots, G_p$, then one can give oneself *an* element $s = (s_1, \ldots, s_p)$ of the set $G_1 \times \cdots \times G_p$, which is also an echelon. If there is a relation $R(x_1, \ldots, x_p)$, one can consider the relation

$$R(x_1, \ldots, x_p) \text{ and } x_1 \in G_1 \text{ and } \ldots \text{ and } x_p \in G_p.$$

Otherwise, one can also take $R$ as a relation of only one $x$ of $G = G_1 \times \ldots \times G_p$.

There is the theorem: $\text{Coll}_x\Big(R(x) \text{ and } x \in G\Big)$, i.e., $R(x)$ determines in $G$ a subset $H \subset G$ such that $\{x \in H \subset R(x) \text{ and } x \in G\}$.

This set $H$ was previously denoted by $E_x(R(x)$ and $x \in G)$. Later we will denote this set $H$ by $\{x \mid x \in G$ and $R(x)\}$ (as already mentioned). The set $H$ is still an element of $\mathcal{P}(G)$, i.e., a relation $R$ can be characterized by a *subset* of an echelon or as an element of an echelon. In the same way, functions, applications, etc. can be characterized by an element of an echelon. In particular the relations of representation $R_\mu$ (see Chap. 3) can thus be described by a subset $r_\mu$ of an echelon $S_\mu$ on the terms of representation as base sets or as elements $r_\mu \in \mathcal{P}(S_\mu)$. One can also naturally consider for all $R_\mu$ the element $(r_1, r_2, \ldots) = s$ of $\mathcal{P}(S_1) \times \mathcal{P}(S_2) \times \cdots$; and if, conversely, $s \in \mathcal{P}(S_1) \times \mathcal{P}(S_2) \times \cdots$, then $s = (r_1, r_2, \ldots)$ is equivalent to

$r_1 \in \mathcal{P}(S_1)$ and $r_2 \in \mathcal{P}(S_2)$ and ….

Axioms or theorems, expressible only by the $R_\mu$, transform themselves into a relation $P$ of $s$ into which enter the base sets.

# 3
# From Reality to Mathematics

In this chapter we will be concerned with the problem of correspondence between the structure of a part of reality and an idealized mathematical structure that is similar to the structure of that part of reality. This part of reality will be called the *application domain* of the theory, denoted by $A_p$.

We will not explain or base this structure on philosophical considerations or other points of view, but by the results of *experiments*. The only foundation of the detected structures is the success of the method of physics that we will describe here in detail. This does not mean that we can detect these structures only by going for a *walk* in nature; it is necessary to make experiments, i.e., to *work* with nature and to observe the behavior of nature. Thus the detected structures can depend on our work with nature. We can explain why we make a particular experiment, but we cannot explain the structure we have verified by the result of this experiment.

It would be a big mistake to think that the realities depend only on the structures given in untouched nature. There are many realities that depend essentially on our labor, e.g., a car produced in a factory. We can explain the many structures of this car because we have produced the car, but we cannot explain the laws of nature that we have used. We can only detect these laws and what is essential to formulate these laws in the form of mathematical structures.

The formulation of this method will be founded in this chapter by describing the transition from reality to mathematics by distinguishing three processes:

- A first process, called *recording process*, which is a formulation of recorded facts denoted under the form of sentences in a natural language of very simple form (a descriptive language) called the basic language of the intended theory. It must be possible to record these facts without using the intended theory.
- A second process, called *mathematization process*, which is a transcription of *natural sentences* denoting facts (formulated in the basic language) into

*formal sentences* (formulated in a formal language) – the mathematical language of the theory.
– A third process, called *idealization process*, which is an enrichment by idealization of the previous mathematical structures of the theory.

## 3.1 Recording Process

The first step in the transition from reality to mathematics is a *recording process* of facts, denoted by $\hookrightarrow$. By this recording process, facts related to the application domain $A_p$ of the theory are recorded under the form of sentences formulated in a natural language of very simple form (a descriptive language) called the basic language of the theory.

In an intuitive way, the presence of such a language ensures us of the existence of a nonproblematic language from a semantic point of view, i.e., a language that has a complete and exclusively empirical interpretation, i.e., using only concepts related to *direct* observations, or *indirect* observations with the help of pre-theories.

### 3.1.1 Basic Language

Despite its potential syntactic–semantic ambiguity and vagueness, natural language still serves as the foundation of any of the formal languages (see Sect. 3.2)[1]. Here we consider a natural language of very simple form, called *basic language*, denoted by $B_l$. In this language, facts of the reality, stating properties of objects and the relations between objects, are denoted by *natural sentences* of the form:

(p) the object $\bar{a}$ has the property $\tilde{p}$;
(r) between the objects $\bar{a}_1, \ldots, \bar{a}_n$ and finite many real numbers $\alpha_1, \ldots, \alpha_n$, there is the relation $\tilde{r}_n(\bar{a}_1, \ldots, \bar{a}_n, \alpha_1, \ldots, \alpha_n)$.

Before we can formulate sentences of the form $(p)$ and $(r)$, it is necessary to establish the *objects*, their *properties* and *relations*. Before we can begin to develop a physical theory, we must have the possibility of such determinations of realities without using the intended new theory. This does not mean that we do not use any theory. On the contrary, for most of the physical theories, we use so-called *pre-theories*, i.e., already known theories. We will discuss later how we can use theories to detect new realities (see Sect. 6.3).

It is clear that we have to begin physics by such theories that do not use pre-theories, i.e., where we can state realities directly by our senses. But we do not want to discuss the problem of a construction of the *total* physics.

---
[1] Later, in Chap. 6, we will extend the basic language $B_l$ to an *extended basic language*, denoted by $B_{l_{ex}}$, by introducing 'new words' designating "new concepts." By these "new concepts" we detect also - new facts -.

We are convinced that the domain of physics is determined by the region of directly observable realities, and by the method of physical theories that we wish to develop in this book. There are many realities which we do not accept as directly observable *physical* realities. This does not mean that we deny the existence of realities that cannot be detected in a physical way. We will hint at such realities in some parts of this book.

The method of physical theories developed in this book can also be used in other domains than simply those of physics if we begin with realities that are not accepted by physics as directly observable realities (see, e.g., [7]). In this way we can, e.g., decide that we are not interested in describing the directly observed sound of a symphony played by an orchestra. The CD of this symphony is an example of a *physical recording* of the vibration of the air produced by the orchestra. To obtain from this CD an impression of the *physically noninteresting* sound of the symphony, which is for us the essential value of the CD, implies the construction of a good hi-fi installation system. However, the construction of this system requires a good physical and psychological knowledge.

The above formulation $(p)$ and $(r)$ of sentences of the basic language is not sufficient; we have to say something more about the possibility that real numbers can appear in sentences of the form $(r)$. They cannot appear from direct observations, since we cannot find real numbers in nature. We can only find whole numbers, e.g., - the number of sheep in a herd -. Therefore real numbers can only appear by the intermediary of pre-theories. By these pre-theories it is established which *objects* are described by real numbers $\alpha$. For instance, if we use Newton's theory of space–time as a pre-theory, we can construct the description of *time* by one real number $t$ as describing a *time-point*, and a triplet of real numbers $(x_1, x_2, x_3)$ as standing for a spot in space. The real numbers $t, x_1, x_2, x_3$ can only be determined by fixing a space–time reference system.

Thus real numbers in $(r)$ only make sense if one says from which pre-theory, and in which way these real numbers appear in $(r)$. But it would be practically impossible to say where a real number appears, and in which way this real number is defined, e.g., a real number $t$ for the "time" ("time" defined in the pre-theory). Therefore one presumes, if we use real numbers in $(r)$, that the definition of these real numbers by pre-theories is known and marks these definitions by only one word as, e.g., 'time $t$' instead of only '$t$', or 'position $x_1, x_2, x_3$' instead of only '$x_1, x_2, x_3$'. Sometimes one removes the words 'time' and 'position', and marks the physical sense by only the *letters*, e.g., '$t$' for time and '$x$' for position. Often the meaning of a real number is also determined by the position of this real number in $(r)$ (see Sect. 3.1.4).

The introduction of real numbers in physics is a very common practice which essentially simplifies the mathematical description, so that one sometimes thinks that the physical method is essentially *quantitative*. But physics is not only quantitative. How real numbers can be defined in physical theories will be described later (see Sect. 3.1.1.2).

A necessary presumption for the possibility of formulating sentences of the forms $(p)$ and $(r)$ is the possibility to mark an - object - by a letter '$\overline{a}_i$' (also by any sign). If we have, e.g., two glass balls, we can mark one glass ball by '$\overline{a}$' and the other by '$\overline{b}$'. This is not trivial. If we have, e.g., introduced by quantum mechanics as a pre-theory the concept of electrons, we see that we cannot mark a particular electron by a letter. If we have, e.g., an He-atom with two electrons, we cannot mark one electron by '$\overline{a}$' and the other by '$\overline{b}$', since the hypothesis that we can distinguish the $\overline{a}$-electron from the $\overline{b}$-electron is in contradiction to quantum mechanics. However, if we make a collision experiment between an electron and an H-atom it would be a contradiction to quantum mechanics to say that the electron leaving the collision is the same (or is not the same) as the incident electron. The electron leaving the collision is neither the incident electron nor that which was in the H-atom.

If we can mark certain realities by a letter, we say that these realities are objects. Electrons are not in this sense objects. A "physical system" consisting of several electrons and nucleus can be an object if it can be identified using the total experiment that will be described by the intended theory.

The fact that we restrict the basic language of an intended theory to sentences of the forms $(p)$ and $(r)$ is essential for the construction of the *method of physics* proposed in this book. But since physicists also use other forms of language that are not very clear and very condensed, so condensed that people other than physicists can (or better, must) misunderstand this language, we will give examples of languages that do not belong to the basic language and that do not describe the correct method of physics.

An example of such a sentence that could be misunderstood is the following:

'A ruler in motion is shorter than a ruler at rest'.

Such a sentence makes no sense. If one analyzes it, it is syntactically correct. Nevertheless, it is semantically incorrect. This sentence is in contradiction with itself.

In this sentence, the first ruler has the property "to be in motion." But what is "in motion"? "In motion" has no "absolute" meaning. It means that the first ruler is moving "relative to the second ruler." Also that, the second ruler is moving "relative to the first ruler." Therefore, the above sentence would also say that the second ruler is shorter than the first ruler – a contradiction.

The above sentence is an incorrect formulation of the following reality: We produce by the first ruler two marked spots *on* the second ruler by the following process: in the middle of the first ruler we produce a flash of light. If this flash reaches the two ends of the ruler, it will produce two marked spots on the second ruler. Then the distance between these two marked spots on the second ruler is smaller than the second ruler.

If we produce in the same way two marked spots on the second ruler by a flash on the second ruler, then these two marked spots on the first ruler also have a distance smaller than the first ruler.

It is clear that such briefly formulated sentences have nothing to do with the intended basic language. But there are many other sentences used by physicists which we do not want to take as sentences of a basic language. We can only formulate correctly the syntax and the semantics of a basic language and provide examples.

A very large area of language is the region of language where physicists speak of fairy tales with the intention of finding a new theory and of finding new realities (see Sect. 6.3). We make no restrictions to such a language, since we make no restrictions to finding intuitively new mathematical parts $MT_\Sigma$ of a physical theory and to finding the picture terms in this $MT_\Sigma$ (see Sect. 4.3). The only interesting question is to know whether this $MT_\Sigma$, with the picture terms, is a good representation of the application domain (see Sect. 4.3). Even unconventional ways are permitted if the result is usable.

In physics there are also linguistic formulations which could be used for a basic language but which are not so suitable for our aim, since they presuppose more or less some structures of reality that we would like to describe by axioms of the mathematical part of the theories.

If we have for instance on the table a glass of water, and if we take as an object $\bar{a}$ this water, we can formulate as a sentence of $B_l$:

'$\bar{a}$ has the property *liquid*'.

There are many equivalent linguistic formulations, e.g.,

'$\bar{a}$ is liquid'   (liquid as an adjective),

'$\bar{a}$ is a liquid'  (liquid as a substantive).

Or another example:

'$\bar{a}$ has the property *gas*'.

There also are other linguistic formulations, e.g.,

'$\bar{a}$ is gaseous',

'$\bar{a}$ is a gas'.

We will not take all these possible equivalent formulations in $B_l$. We will only take the formulation '$\bar{a}$ has the property *liquid*' or '$\bar{a}$ has the property *gas*' as sentences of $B_l$. It appears to us that it brings essential advantages if one allows only restricted forms of sentences in $B_l$.

Another example: We have measured the distance value $\alpha$ between two objects with the help of a pre-theory of measuring distances. The physicists use for the majority of time very simple, and therefore also very imprecise, formulations, as for example,

'The distance between the objects $\bar{a}_1$ and $\bar{a}_2$ has the value $\alpha$'.

Such a sentence is not correct since there is *no exact* real number $\alpha$ as a value of the distance! Therefore we want to allow in $B_l$ only a correct formulation as, e.g.:

'Between the objects $\bar{a}_1, \bar{a}_2$ and the real number $\alpha$, there is the distance relation $\delta(\bar{a}_1, \bar{a}_2, \alpha)$'.

There can be different $\alpha_1, \alpha_2, \ldots$ which fulfill this relation!

Therefore it is advantageous if we allow in $B_l$ only a very restricted form of sentences. This does not mean that we do not allow physicists to speak amongst themselves in a very vague language, so that people who are not "insiders" can (or even must) misunderstand this language.

### 3.1.1.1 The Syntax of the Basic Language

The syntax of the basic language $B_l$ is given by the following postulates:

(i) the sentences are of the forms $(p)$ and $(r)$,
(ii) we can form *compound natural sentences* made up of sentences by using the conjunctive word 'and',
(iii) we can take the *negation of a sentence* of the forms $(p)$ or $(r)$ by using the word 'not'.

We call (ii) and (iii) the logic of the basic language. It is essential that we do not allow words such as 'all' and 'there exists' (or 'there are'). The logic of $B_l$ is therefore a very simple one. We will see later that if sentences designate propositions that have a meaning, then some of these propositions can be additionally assigned a truth value (see Sect. 3.1.1.2). But the logic of $B_l$ has nothing to do with this notion of "truth"!

We therefore do not introduce a double negation as, e.g., 'not : $\bar{a}$ has not the property $\widetilde{p}$', and therefore we do not also introduce the "axiom" that this double negation is equivalent to the sentence '$\bar{a}$ has the property $\widetilde{p}$'.

We also do not introduce the negation of a compound natural sentence composed of sentences $(p)$ and $(r)$.

The logic of $B_l$ is therefore very weak; it will be fixed later by the translation of $B_l$ into a mathematical language and by the axioms for the mathematical logical signs '$\wedge$' and '$\neg$' (see Sects. 2.2 and 3.2.1).

### 3.1.1.2 The Semantics of the Basic Language

The meaning of the words in a sentence are given by a more or less long process. This process starts with a meaning of words that we introduce without a theory, i.e., we introduce words by simple immediate demonstrations as in the following sentences: 'this is a *desk surface*', 'this is a *stone*', 'this is a *liquid*', 'this is a *rigid body*', etc.

We will not describe what we allow as having directly demonstrable meaning. Physics is essentially determined by the domain which we admit as having directly demonstrable meaning. For example, we do not allow a direct meaning to words such as 'weak', 'blue', or 'weak sound'.

But in physics there are many words the meanings of which is given by pre-theories (see sect. 5.3). For example, the meaning of the word 'distance' is given by a pre-theory. The words used in a basic language must be words the meanings of which *are* given without the intended theory.

The semantics of the basic language $B_l$ is given by assigning *semantic relations* between linguistic, conceptual, and real entities. The schema given in Fig. 3.1 summarizes these semantic relations.

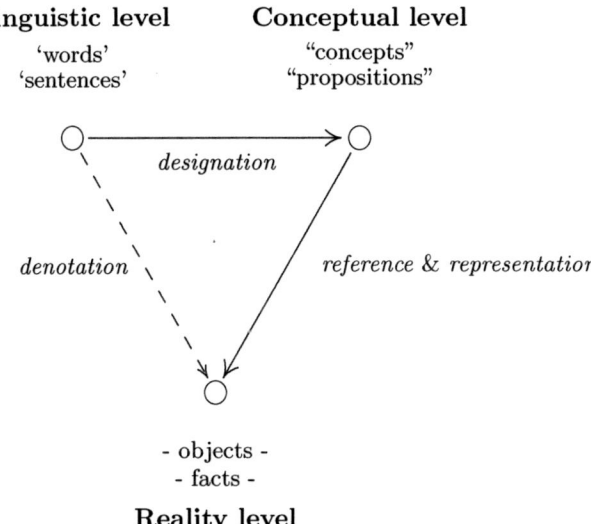

**Fig. 3.1.** Semantic relations

One can understand that linguistic entities designate conceptual entities. But how we get relations between conceptual entities and real entities is a very complicated process: We will not describe the process of obtaining relations by "immediate reference;" we can only say which of these concepts are used, but how we get such concepts is a psychological process which we cannot avoid by introducing philosophical a priori concepts.

Today we try to reduce all facts described by the basic language $B_l$ to "digital" facts (not necessarily observed immediately!). These digital facts have the advantage in that the "negation" of a digital fact is also a digital fact, so that there are no problems by using the usual logical axioms for the words 'not' and 'and'. If we have, e.g., a nondigital property $\widetilde{p}$, it can be that we cannot decide whether '$\overline{a}$ has the property $\widetilde{p}$' or '$\overline{a}$ has not the property $\widetilde{p}$', i.e., that we must leave open the fact that '$\overline{a}$ has or has not the property $\widetilde{p}$'.

The extended language $B_{l_{ex}}$ has the same form as $B_l$ but with the additional 'new words' related to these "new concepts" and - new detected realities -. Therefore we will try to show in a general form the connections between the linguistic, the conceptual, and the reality levels.

The connection between the linguistic level and the conceptual level is very important, especially for the extended language $B_{l_{ex}}$, since one often uses words the meaning of which is already known, but which have a totally different meaning in physics. For instance, the word 'atom' has (in the Greek language) the meaning "indivisible;" but an "atom" in physics (i.e., in a $B_{l_{ex}}$) is divisible in many parts. Therefore, one can make the joke that "an atom is called atom (i.e., indivisible) because it is not an atom (i.e., it is divisible)." But there are also many other words in physics (i.e., in a $B_{l_{ex}}$) which are new and which also have a new meaning (i.e., which stand for a new concept). How we can detect new realities will be described in detail in Sect. 6.3.

The following descriptions of the connections between the three semantic levels is also important as a predevelopment for the purpose of recording the results of physical experiments (the recorded facts) in a database, i.e., how we can collect the various $\widetilde{A}$ (introduced in Sect. 3.1.4) as part of this database.

*Notes concerning the Sect. 3.1.1.2*

We will distinguish real, linguistic, and conceptual entities by enclosing the real entities in dashes, e.g., the fact - the object $\overline{a}$ has the property $\widetilde{p}$ -; the linguistic entities in single quotes, e.g., the sentence 'the object $\overline{a}$ has the property $\widetilde{p}$'; the conceptual entities in double quotes, e.g., the proposition "the object $\overline{a}$ has the property $\widetilde{p}$." We will enclose the semantic meta-propositions in corners, e.g., ⌈ the sentence '$s$' designates the proposition "$p$" ⌉.

*Basic Properties*

Some properties called *basic properties* are selected. This selection is not imposed by the reality but is selected by our own free will; it is the choice of deciding for which region of the reality we intend to make a theory. We want to exclude from the application domain $A_p$ all objects that do not have at least one basic property!

*Example A*

Throughout Part I, the introduced concepts will be illustrated on the basis of a simple example, noted *Example A*, "*A description of the surface of the earth, or of a round table.*" In this context, we consider the following:

- Only *one* property concept "*marked spot*" denoted by $\tilde{p}$. In the basic language $B_l$ we will formulate sentences such as 'the object $\bar{a}$ has the property *marked spot*'.
- Only *one* relation concept "*distance relation between objects*" denoted by $\tilde{r}$. In the basic language $B_l$ we will formulate sentences such as 'between the objects $\bar{a}_1, \bar{a}_2$ and the real number $\alpha$, there is the *distance relation* $\delta(\bar{a}_1, \bar{a}_2, \alpha)$', where the *distance relation* $\delta(\bar{a}_1, \bar{a}_2, \alpha)$ is obtained on the basis of a pre-theory of the measurement of distance.

We *decide* that the property concept "*marked spot*" is to be taken as a "basic property." This means that we want to describe by the intended theory only such objects that possess the property *marked spot*.

**Relation of Designation**

The *relation of designation* holds between some members at the linguistic level and their correlates at the conceptual level. This relation of designation occurs in meta-propositions such as

⌈ the sentence '$s$' designates the proposition "$p$" ⌉.

We stipulate that linguistic meaning is a property that a linguistic entity (i.e., word, sentence) acquires when it happens to designate a conceptual entity (i.e., concept, proposition), in other words, an entity of significance. We will see that the relation of designation (i.e., the relation taking entities of significance as values) composed with the conceptual meaning provides the linguistic meaning.

We assume the principle of *semantic compositionality*. Roughly speaking, this principle says for a sentence that

*the meaning of the whole depends on the meaning of its parts.*

According to this principle, the meaning of the sentence depends on the meaning of the words. In the recording process, the sentences formulated in the basic language $B_l$ can be considered from a syntactic and semantic point of view as free from ambiguity. In this case, we admit that

*the designation of the whole depends on the designation of its parts.*

Nevertheless, the acceptance of this principle does not mean that we adopt the extreme thesis of semantic atomism. Our view is rather contextual.

## Example A

Considering the sentence as a whole:

- ⌈ the sentence '$\bar{a}$ has the property marked spot' designates the proposition "$\bar{a}$ has the property marked spot" ⌉,
- ⌈ the sentence 'between the objects $\bar{a}_1, \bar{a}_2$ and the real number $\alpha$, there is the distance relation $\delta(\bar{a}_1, \bar{a}_2, \alpha)$' designates the proposition "between the objects ..." ⌉.

### Relation of Reference

The *relation of reference* holds between members at the conceptual level and their correlates at the reality level. This relation of reference occurs in meta-propositions such as

⌈ the proposition "$p$" refers to - the object $o$ - ⌉.

## Example A

- ⌈ the proposition "$\bar{a}$ has the property marked spot" refers to - the object $\bar{a}$ - ⌉,
- ⌈ the proposition "between the objects $\bar{a}_1, \bar{a}_2$ and the real number $\alpha$, there is the distance relation $\delta(\bar{a}_1, \bar{a}_2, \alpha)$" refers to - the objects $\bar{a}_1, \bar{a}_2, \alpha$ - ⌉.

If there appear real numbers $\alpha_i$ in a relation $\tilde{r}(\bar{a}_1, \ldots, \alpha_1, \ldots)$, then they refer to objects that are defined in the context of a pre-theory, and can be described by real numbers that are also defined in a pre-theory. For example, they can refer to marked spots and time-points that are defined in the context of a pre-theory of Newton's space–time reference system, and can be described by real numbers $t$ for time and three rectangular coordinates $(x_1, x_2, x_3)$ which are also defined in a pre-theory (See Sect. 3.1.3).

### Relation of Representation

The *relation of representation* holds between some members at the conceptual level and their correlates at the reality level. This relation of representation occurs in meta-propositions such as

⌈ the proposition "$p$" represents the fact - $f$ - ⌉.

## Example A

- ⌈ the proposition "$\bar{a}$ has the property marked spot" represents the fact - $\bar{a}$ has the property marked spot - ⌉,
- ⌈ the proposition "between the objects $\bar{a}_1, \bar{a}_2$ and the real number $\alpha$, there is the distance relation $\delta(\bar{a}_1, \bar{a}_2, \alpha)$" represents the fact - between the objects ... - ⌉.

The relations of reference pairs "propositions" to - objects - (each object as a whole); the relations of representation pairs "propositions" to - aspects of objects - (i.e., - properties of objects - or - relations between objects -). The schema given in Fig. 3.2 summarizes these relations.

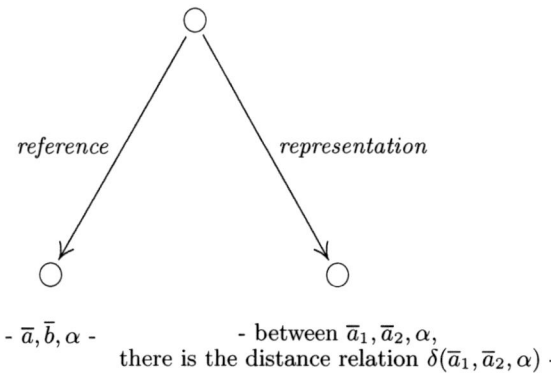

Fig. 3.2. Semantic relations of reference and representation

An adequate representation of a whole domain of facts requires a system of concepts and propositions.

### Relation of Denotation

The *relation of denotation* holds between some members at the linguistic level and their correlates at the reality level. This relation of denotation is construed as the composition of the relations of designation, and of reference or representation between the 'linguistic level' and the - reality level - via the "conceptual level"!

### Truth of a Proposition

Whether a proposition "$p$" is true can be decided by a process that the physicists call "observation." This process can be of "direct observations," i.e., by observations that have been made without any theory (e.g., a cup of tea on a table) and of "indirect observations," i.e., by using a pre-theory by which we decide whether a proposition represents a fact of reality or not. This decision is made by the methods described in Chap. 6, and which are applied to the *pre-theory*. This process also tells us the "imprecision of the observation," i.e., that

- ⌈ the proposition "$p$" is true only with a certain imprecision ⌉,

i.e., that

- ⌈ not the proposition "$p$" is true, but another proposition "$\widehat{p}$" ⌉,

which one gets in $B_l$ by introducing the imprecision, e.g., by replacing a real number $\alpha$ by an interval $J$ of real numbers. The interval $J$ describes the "error of measurement."

### 3.1.2 Application Domain of a *PT*

The *application domain* of a physical theory $PT$, denoted by $A_p$, is the domain of facts (stating properties and relations between objects) which is considered in the recording process.

At the linguistic level, this domain of facts is denoted under the form of a collection of sentences formulated in the basic language $B_l$. At the conceptual level, this domain of facts is designated under the form of a collection of propositions. This means that only the facts denoted by sentences, using terms that designate property or relation concepts belonging to the context related to the application domain $A_p$, can be taken into account.

Let us recall that we want to exclude from $A_p$ all objects that do not have at least one basic property!

*Example A*

We decide that the property concept "marked spot" is to be taken as a basic property, i.e., that we want to describe by the intended theory only such objects which are marked spots. The application domain $A_p$ consists of - marked spots - and - distance relations between marked spots -.

### 3.1.3 Recording Rules

By *recording rules*, we mean rules that ensure us the correspondence between a fact and its formulation under the form of a sentence in the basic language $B_l$.

These recording rules are not made directly. The recording rules are in fact the relations of denotation described in Sect. 3.1.1. These relations of denotation are construed as the composition of the relations of designation and of reference between the 'linguistic level' and the - reality level - via the "conceptual level"!

### 3.1.4 Facts Recorded in the Basic Language

The result of the recording process of recorded facts is denoted under the form of a *finite collection of natural sentences*, also called a *compound natural sentence* or simply *text* (the term generally used hereafter!) denoted by $\widetilde{A}$ formulated in the basic language $B_l$. Here, it should be distinguished between recorded facts: (i) *stated* (the factual truth is stated) and (ii) *not stated* or *not knowing that they were stated*.

*Example A*

We have a finite collection of sentences such as

'the object $\bar{a}_1$ has the property marked spot *and* the object $\bar{a}_2$ has the property marked spot *and* between $\bar{a}_1, \bar{a}_2$ and the real number $\alpha \in J_1$ there is the distance relation $\delta(\bar{a}_1, \bar{a}_2, \alpha)$'.

The possibility to use in $B_l$ a mathematical relation (i.e., $\alpha \in J_1$) is due to the fact that we use pre-theories. In this way we replace the distance relation '$\delta(\bar{a}_1, \bar{a}_2, \alpha)$' by the distance relation

'$\delta(\bar{a}_1, \bar{a}_2, J)$ : there is a real number $\alpha \in J$ with $\delta(\bar{a}_1, \bar{a}_2, \alpha)$'.

In a general manner, we have a finite collection of sentences of the form

(p) 'the object $\bar{a}$ has the property $\widetilde{p}$';
(r) 'between the objects $\bar{a}_1, \ldots, \bar{a}_n$ and finite many real numbers $\alpha_1, \ldots, \alpha_n$, there is the relation $\widetilde{r}_n(\bar{a}_1, \ldots, \bar{a}_n, \alpha_1, \ldots, \alpha_n)$'.

We have in Sect. 3.1 used the real number $\alpha$ as the symbol of a possible result of measurement with the help of a pre-theory.

Now it is necessary to say a little bit more about the description of measurements by pre-theories.

In general, we can have more than one kind of measurement. Therefore, there can be more than one real number in $\widetilde{r}(\bar{a}_1, \ldots, \alpha_1, \ldots)$. The choice of pre-theories establishes from which physical reality the various $\alpha$ are measurements. For instance, for a gas in a container we can measure the *pressure* and the *volume*. By the *position* $\alpha_1$ as the first and the *position* $\alpha_2$ as the second in $\widetilde{r}$ it will be fixed that $\alpha_1$ *is* the pressure and $\alpha_2$ *is* the volume.

It can also be that we have for the description of the results of measurement not only one real number, but a collection of results of measurement, e.g., $\overline{M} = \overline{\mathbb{R}} \times \overline{\mathbb{R}} \times \overline{\mathbb{R}}$ for the measurements of the orthogonal coordinates of the position of an object. The set $\overline{M}$ must be defined by the pre-theories as a set with a uniform structure (see Sect. 4.6) by the elements of which can be described the errors of measurement.

In order to simplify our description, we will symbolize the results of measurement only by $\overline{\mathbb{R}}$ and intervals $J$ for the description of the errors of measurement.

It will be mentioned that the errors of measurement, described by intervals, are not necessarily the same for all relations and measurements that appear in the text $\widetilde{A}$!

The next step from reality to mathematics will be to transcribe these *natural* sentences into *formal* sentences, in particular, to transcribe the notion of a collection of sentences of $B_l$ into the notion of a (mathematical) set.

## 3.2 Mathematization Process

The second step in the transition from reality to mathematics is a *mathematization process*, denoted by (cor). By this process, natural sentences formulated in the basic language $B_l$ (related to the application domain $A_p$) will be transcribed into formal sentences formulated in a formal language, the mathematical language of the theory.

### 3.2.1 The Basic Mathematical Theory

In this paragraph we will describe a formal language, i.e., a mathematical language in which it is possible to transcribe natural sentences formulated in the basic language $B_l$, i.e., natural sentences denoting facts of the application domain $A_p$.

As a starting mathematical language, we consider the mathematical language $MT$ described in Chap. 2, i.e., "well-formed" assemblies of signs, logic, and set theory. The mathematical language $MT$ can further contain some constants and axioms for these constants (but not necessarily!). In order to obtain a more comprehensive mathematical language responding to our needs we proceed as follows:

(i) We add to $MT$ two relations: for every property $\widetilde{p}$ of $B_l$, a formal relation $\overline{p}(x)$; and for every relation $\widetilde{r}$ of $B_l$, a formal relation $\overline{r}(x_1, \ldots, x_n, \alpha)$. The relations $\overline{p}, \overline{r}$ are new constants added to $MT$. (If a relation $\widetilde{r}$ contains real numbers (derived from pre-theories), the corresponding $\overline{r}$ will contain the same real numbers.)

With this enrichment,

- the natural sentence of the basic language $B_l$ 'the object $\overline{a}$ has the property $\widetilde{p}$' corresponds to the formal sentence of the enriched $MT$ '$\overline{p}(\overline{a})$';
- the natural sentence of the basic language $B_l$ 'between the objects $\overline{a}_1, \overline{a}_2, \ldots,$ and the number $\alpha$, there is the relation $\widetilde{r}(\overline{a}_1, \overline{a}_2, \ldots, \alpha)$' corresponds to the formal sentence of the enriched $MT$ '$\overline{r}(\overline{a}_1, \overline{a}_2, \ldots, \alpha)$'.

## 3.2 Mathematization Process

In a similar way, the natural sentences compounded by the connective words 'and' and 'not' in $B_l$ correspond to the formal sentences compounded by the logical signs '∧' and '¬' in the enriched $MT$.

Thus we get a one-to-one correspondence between the natural sentences in $B_l$ and the formal sentences in the enriched $MT$. We also say that the constants $\bar{p}, \bar{r}$ and the logical signs '∧' and '¬' are *interpreted* by the mathematization process (cor).

(ii) We add to $MT$ not only constants but also two axioms for these constants. In $B_l$ some properties are characterized as basic properties. Let $\tilde{p}_1, \tilde{p}_2, \tilde{p}_3$ be the basic properties in $B_l$, and let $\bar{p}_1, \bar{p}_2, \bar{p}_3$ be the corresponding relations added by (i) to $MT$. We then define the relation $\bar{p}_0 = \bar{p}_1 \vee \bar{p}_2 \vee \bar{p}_3$.

The first axiom is the *collectivizing axiom*

$$\text{Coll}_x \bar{p}_0(x). \tag{3.2.1}$$

The relation

$$\exists y (\bar{p}_0(x) \Leftrightarrow x \in y)$$

and, briefly,

$$\bar{p}_0(x) \Leftrightarrow x \in y$$

are equivalent. The set $y$ is then uniquely determined by $\bar{p}_0(x)$; we write $y = \{x \mid \bar{p}_0(x)\}$. Instead of $\bar{p}_1(x) \vee \bar{p}_2(x) \vee \bar{p}_3(x)$ we can write $x \in y$.

The second axiom is

$$\{x \mid \bar{p}_0(x)\} \text{ is a finite set.} \tag{3.2.2}$$

We can never have a contradiction to this axiom (3.2.2), since we can record only finitely many facts (stating "basic" properties, or other properties, of objects and relations between objects), i.e., there can be only finitely many objects in the text $\tilde{A}$. In some cases, we can "prove" this axiom (3.2.2) by an experiment, e.g., if $y$ is the set corresponding to a herd of sheep, one can then count the elements of $y$. But in physics there are sets in which the number of elements is so large that we can never count all of these elements since we can never record "all" of the objects.

In conclusion, we take together the constants $\bar{p}, \bar{r}$ and the axioms (3.2.1) and (3.2.2) under the sign $\widehat{\Theta}$. For the theory enriched in this way we write $MT_{\widehat{\Theta}}$.

*Example A*

We add to $MT$ two new constants $\bar{p}$ and $\bar{r}$: $\bar{p}$ is a relation of weight 1 noted $\bar{p}(x)$, and $\bar{r}$ is a relation of weight 3 noted $\bar{r}(x_1, x_2, \alpha)$ where $\alpha$ is a real number $\alpha \in \overline{\mathbb{R}}$ and $\overline{\mathbb{R}}$ is a finite set of real numbers. We have taken $\tilde{p}$ as a basic property. Therefore, on the basis of the axiom (3.2.1), there is a set $\overline{M}$ with

$$x \in \overline{M} \Leftrightarrow \bar{p}(x).$$

On the basis of the axiom (3.2.2), $\overline{M}$ is a finite set.

(We have introduced $MT_{\widehat{\Theta}}$ only in order to show that the "collectivizing" axiom is the basis of the standard form $MT_\Theta$, which we will use in all future descriptions of physical theories.)

### 3.2.2 The Standard Mathematical Theory

It is very convenient on mathematical grounds to change a little the definition of $\widehat{\Theta}$ to $\Theta$. Instead of introducing the constants $\bar{p}, \bar{r}$ and the axioms (3.2.1) and (3.2.2), we introduce as $\Theta$ constants $\overline{M}_0$, several $\bar{s}$, and the axioms:

$$\overline{M}_0 \text{ is a finite set;} \qquad (3.2.3)$$

and for the various $\bar{s}$,

$$\bar{s} \subset \overline{M}_0 \qquad (3.2.4)$$

or

$$\bar{s} \subset \overline{M}_0 \times \overline{M}_0 \times \cdots \times \overline{\mathbb{R}} \times \cdots, \qquad (3.2.5)$$

where $\overline{\mathbb{R}}$ is a *finite* set of real numbers.

We define the following correspondences between $MT_{\widehat{\Theta}}$ and $MT_\Theta$:

$$\{x \mid \bar{p}_0(x)\} \text{ corresponds to } \overline{M}_0, \qquad (3.2.6)$$

$\bar{p}(\bar{a})$ (for a property $\bar{p}$) *corresponds to* $\bar{a} \in \bar{s}$
(for the $\bar{p}$ corresponding to $\bar{s}$), $\qquad (3.2.7)$

$\bar{r}(\bar{a}_1, \bar{a}_2, \ldots, \alpha)$ (for a relation $\bar{r}$) *corresponds to* $(\bar{a}_1, \bar{a}_2, \ldots, \alpha) \in \bar{s}$
(for the $\bar{r}$ corresponding to $\bar{s}$). $\qquad (3.2.8)$

Since we have postulated that all objects $\bar{a}$ of the application domain $A_p$ have the property $\bar{p}_0$, we can add this property $\bar{p}_0$ to every $\bar{p}(x)$ and to every $\bar{r}(\bar{a}_1, \bar{a}_2, \ldots, \alpha)$:

"$\bar{p}(x)$ and $\bar{p}_0(x)$;"

and

"$\bar{r}(\bar{a}_1, \bar{a}_2, \ldots, \alpha)$ and $\bar{p}_0(x_1)$ and $\bar{p}_0(x_2)$ and ...."

The above correspondences (3.2.6) to (3.2.8) follow then from the theorems (see Sect. 2.3)

"$\bar{p}(\bar{a})$ and $\bar{p}_0(\bar{a})$" $\Leftrightarrow$ $\bar{a} \in \{x \mid \bar{p}(x)\}$;

and

"$\bar{r}(\bar{a}_1, \bar{a}_2, \ldots, \alpha)$ and $\bar{a}_i \in \{x \mid \bar{p}_0(x)\}$" $\Leftrightarrow$

$(\bar{a}_1, \bar{a}_2, \ldots, \alpha) \in \{(x_1, x_2, \ldots, \alpha) \mid \bar{r}(x_1, x_2, \ldots, \alpha)\}$.

Together with the correspondence between the natural sentences of $B_l$ and the formal sentences of $MT_{\hat{\Theta}}$ (see Sect. 3.2.1) we get a correspondence between the natural sentences of $B_l$ and the formal sentences of $MT_\Theta$. For this correspondence we write also $B_l$ (cor) $MT_\Theta$.

It is often more convenient not to introduce $\bar{p}_0(x)$ and the corresponding $\overline{M}_0$ but to take all basic properties and to introduce for each basic property a base set $\overline{M}_i$. Thus, between the natural sentences of $B_l$ and the formal sentences of $MT_\Theta$ we have the following correspondences:

- 'the object $\bar{a}$ has the basic property $\widetilde{p}_i$' *corresponds to* '$\bar{a} \in \overline{M}_i$';
- 'the object $\bar{a}$ which has the basic property $\widetilde{p}_i$ also has the property $\widetilde{p}$'(basic or not) *corresponds to* '$\bar{a} \in \bar{s} \subset \overline{M}_i$' (with $\bar{s}$ corresponding to "$\widetilde{p}_i$ and $\widetilde{p}$");
- 'between the object $\bar{a}_1$, which has the basic property $\widetilde{p}_1$, and the object $\bar{a}_2$, which has the basic property $\widetilde{p}_2$, there is the relation $\widetilde{r}_n(\bar{a}_1, \bar{a}_2, \cdots)$' *corresponds to* '$(\bar{a}_1, \bar{a}_2, \ldots) \in \bar{s} \subset \overline{M}_{i_1} \times \overline{M}_{i_2} \times \cdots$'

Thus the mathematization process $B_l$(cor)$MT_\Theta$ is well defined also for this general case where we take all basic properties separately.

As an axiom we only postulate that

all $\overline{M}_i$ are finite sets.

*Example A*

We add to $MT$ two new constants $\overline{M}$ and $\bar{s}$. On the basis of the axiom (3.2.3), $\overline{M}$ is a finite set. On the basis of the axiom (3.2.5), we have

$\bar{s} \subset \overline{M} \times \overline{M} \times \mathbb{R}$.

The mathematization process $B_l$ (cor) $MT_\Theta$, i.e., the transcription of natural sentences formulated in the basic language into formal sentences formulated in the formal language, is given by

'$\bar{a}$ is a marked spot' (cor) '$\bar{a} \in \overline{M}$',

'the measured distance between $\bar{a}$ and $\bar{b}$ is $\alpha \pm \varepsilon$' (cor) '$(\bar{a}, \bar{b}, J) \cap \bar{s} \neq \emptyset$'.

Here $J$ is the interval $\alpha - \varepsilon$ to $\alpha + \varepsilon$.

### 3.2.3 Enrichment of $MT_\Theta$ by $\overline{A}$

By the mathematization process (cor) we have established a correspondence between a *finite collection of natural sentences* (also called *compound natural sentence* or simply *text*) denoted by $\widetilde{A}$, formulated in the basic language $B_l$, and a *finite collection of formal sentences* (also called *compound formal sentence* or simply *text*) denoted by $\overline{A}$, formulated in the mathematical language $MT_\Theta$.

Here, we make the distinction between recorded facts: (i) "stated" (the factual truth is stated), and (ii) "not stated" or "*not knowing that they were stated.*" In Chap. 6 we will give an answer to the question: How can we "state" real facts with the help of a $PT$ even if they were not stated before by direct observations or with the help of pre-theories? The text $\overline{A}$ will only contain sentences denoting facts that were stated by direct observations or with the help of pre-theories. Often $\overline{A}$ does not contain "all" sentences denoting "stated facts" but only a part of them.

The reason for which we can use mathematics in physics depends on the fact that in physics we consider only facts that can be denoted in a simple basic language $B_l$ (see Sect. 3.1.1).

If we add to $MT_\Theta$ the results of a mathematization process, i.e., the text $\overline{A}$, then we obtain an enriched theory, denoted by $MT_\Theta \overline{A}$. The objective signs $\bar{a}$ of $\overline{A}$ are to take as constants in the theory $MT_\Theta \overline{A}$. In practice, we will not take in $\overline{A}$ "all" results of experiments made until now, but only a part of these experiments – also different parts as different texts $\overline{A}_\eta$. Most of the time, we have *only in mind* that one takes together these parts as one text $\overline{A}$. The formal sentences of $\overline{A}$ are of the form

$$\bar{a} \in \overline{M}$$

and (3.2.9)

$$(\bar{a}_1, \bar{a}_2, \ldots, \alpha) \in \bar{s}, \quad \text{for an} \quad \alpha \in J,$$

where the interval $J$ describes the "error of measurement." This can also be written in the form

$$(\bar{a}_1, \bar{a}_2, \ldots, J) \cap \bar{s} \neq \emptyset, \tag{3.2.10}$$

where $(\bar{a}_1, \bar{a}_2, \ldots, J)$ is the set $\{(\bar{a}_1, \bar{a}_2, \ldots, \alpha) \mid \alpha \in J\}$. Sometimes it can also be interesting to add, as an experimental result, relations of the form

$$(\bar{a}_1, \bar{a}_2, \ldots, J) \subset \bar{s}\,'. \tag{3.2.11}$$

If $\bar{s} \subset \overline{M}_i \times \cdots \times \overline{M}_j$ without any real numbers, then (3.2.10) takes the form

$$(\bar{a}_1, \bar{a}_2) \in \bar{s}$$

and (3.2.11) the form

$$(\bar{a}_1, \bar{a}_2) \in \bar{s}\,'.$$

There can never be a contradiction in $MT_\Theta \overline{A}$ if we have not made mistakes by recording facts or by writing down $\overline{A}$. This follows from the structure of $\overline{A}$ since we have only in $\overline{A}$ relations of the form $x \in y$ (for finitely many $x$ and $y$) without presuming any other axioms except that "all sets are finite sets."

The theory $MT_\Theta \overline{A}$ has a certain analogy with the database of an information system in computer science. Indeed $MT_\Theta$ is similar to the recording structure of the database and $\overline{A}$ is similar to the data stored according to that recording structure.

The theory $MT_\Theta$ is not a physical theory which can be tested by $\overline{A}$. Until now we did not introduce something that we call "physical laws," which tell us something about the structure of the results of experiments, i.e., about the structure of $\overline{A}$. We try to formulate these laws by axioms for the constants introduced by $\Theta$. This method will be described in Sect. 3.3.

*Example A*

We add to $MT_\Theta$ new constants $\bar{a}_1, \bar{a}_2, \ldots$ for some *recorded* marked spots. As "axioms" we add the recorded and, by (cor), transcribed facts, i.e., sentences or relations noted by $\overline{A}$:

$$\bar{a}_i \in \overline{M}\,, \ \ldots$$

and

$$(\bar{a}_i, \bar{a}_k, J_{ik}) \cap \bar{s} \neq \emptyset\,, \ \ldots$$

Since we have in $MT_\Theta$ not introduced any axiom for the subset $\bar{s}$ of $\overline{M} \times \overline{M} \times \mathbb{R}$, the theory $MT_\Theta \overline{A}$ cannot lead to contradictions if we have not made recording errors.

52    3 From Reality to Mathematics

We did not introduce, for instance, the axiom that $\bar{s}$ determines a mapping $\overline{M} \times \overline{M} \to \overline{\mathbb{R}}$! We do not believe that there is one and only one real number determining the distance between two marked spots.

### 3.2.4 The Finiteness of Physics

In the context of an application domain $A_p$, only a finite number of observations can be made, i.e., only a finite number of facts can be recorded. It follows that the collections of sentences of $\widetilde{A}$ and also of $\overline{A}$ are finite. Thus in $\overline{A}$ there can appear only finitely many elements of the set $\overline{M}_i$, which are constants of $\Theta$. No observation can establish that these sets $\overline{M}_i$ have more than finitely many elements. Therefore, observations can never refute an axiom that these sets are finite.

One could perhaps object that in such cases, where we describe the measured values by real numbers, it would be necessary to take the infinite set of real numbers. But this is not necessary since these values are given (on the basis of pre-theories) not by one precise number, but by an imprecise number; for example, two numbers $\alpha$ and $\alpha + 10^{-(10^{(10^{10})})}$ can never be distinguished.

Therefore, it suffices to take a finite set of real numbers to describe all observations.

If we use in $\Theta$ only finite sets (also finite sets of real numbers) this can never be refuted by observations.

But we want to describe by $\Theta$ not only the recorded facts, but also the nonrecorded facts, i.e., the total application domain $A_p$. The finitely established number of objects do not prove that in reality there are only finitely many objects. There are physicists who believe that the fact that all important physical theories use mathematics with infinite sets implies that reality is infinite.

The authors of this book have a completely different belief:

> *The infinity of the sets used in all important physical theories does not come from the reality which is described by theories, it comes from nonrealistic idealizations introduced to simplify the theories. All physical theories using infinite sets are incorrect if one takes these theories as "exact" descriptions of reality. They are only "similar" to the reality. This similarity will be described in Sect. 3.3 by introducing $\Delta$.*

There are some events in the development of physics which do not prove, but merely point out, the finiteness of reality that can be described by physics.

In the beginning, Newton's mechanics described the real space by Euclidean geometry: infinite in the large and in the small; time was also described by an infinite set. The velocity of the masses in Newton's mechanics has no finite boundary. It seems that Newton already doubted his theory for long distances.

But it was found that velocity has a *finite* upper bound described by the special relativity theory. Later on, by Einstein's gravitational theory it was possible to describe finite space and finite time in the large. In the small, space and time were left infinite.

Quantum mechanics refuted the belief that there is an infinite chain of precision. In fact, quantum mechanics can be founded on one axiom of finiteness as explained, e.g., in [8].

The idealization of infinity in the small of space–time (represented by a so-called differentiable manifold) seems to be much deeper than it appears, so that until now it has been impossible to formulate a quantum field theory. In quantum mechanics, an infinity in the small of space–time corresponds to unbounded values of energy in the large, a fact that appears wrong when compared to reality.

All these arguments lead to our axiom in $\Theta$ that the introduced sets are finite, an axiom that can never be refuted by observations. This axiom will have consequences later also in relation to the sets used in the idealized theories. But this can be understood only after introducing a new mathematical concept: the concept of species of structures.

## 3.3 Idealization Process

We know that practically all mathematical theories used in physical theories can only be approximations of reality, i.e., they can be applied to an application domain $A_p$ only under the assumption of allowing for some *degree of approximation* or *degree of inaccuracy*. In this section, we shall proceed to the enrichment of $MT_\Theta$ by idealization.

### 3.3.1 Transition from $MT_\Theta$ to $MT_\Delta$

The system $\Theta$ of constants added to $MT$ (to get $MT_\Theta$) is characterized by base sets $\overline{M}_i$ and by subsets $\overline{s} \subset \overline{M}_{i_1} \times \cdots \times \mathbb{R} \times \cdots$ (see Scct. 3.2.2).

The system $\Delta$ is characterized by the same constants as $\Theta$ and by the following procedure:

- For every base set $\overline{M}_i$ is defined in $MT$ a sct $Q_i$,
- For every $\overline{s} \subset \overline{M}_i \times \cdots$ is defined in $MT$ a subset $s \subset Q_{i_1} \times \cdots \times \mathbb{R} \times \cdots$,
- For every $Q_i$ and every $\overline{Q}_i \times \cdots \times \mathbb{R} \times \cdots$ with $s \subset Q_i \times \cdots \times \mathbb{R} \times \cdots$ are defined in $MT$ "inaccuracy sets" $U_i$ respectively $U_s$.

As "inaccuracy set $U$ for a set $X$" is denoted a set $U \subset X \times X$ which comprises the diagonal set $D$ of all pairs $(x, x)$, for which the relation $(x, y) \in U \Rightarrow (y, x) \in U$ is a theorem, and for which the later relation (3.3.4) is valid.

We now start to get the axiom for $\Delta$. By this axiom we want to postulate that the $Q_i, s$ are "idealized" picture sets of the "physical" sets $\overline{M}_i, \overline{s}$ of $\Theta$. Therefore, we consider mappings

$$\phi_i : \overline{M}_i \to Q_i. \tag{3.3.1}$$

All mappings of product spaces $\overline{M}_i \times \cdots \times \mathbb{R} \times \cdots$ onto $Q_i \times \cdots \times \mathbb{R} \times \cdots$ which are generated canonically by the $\phi_i$ and by the identical mapping $\mathbb{R} \to \mathbb{R}$ will be denoted by $\phi$.

If the $\phi_i$ were bijective and if for $U$ the diagonal sets are chosen, then $\overline{M}_i$ with the $\overline{s}$ would be isomorphic to the $Q_i$ with the $s$. In practically all physical theories, bijective mappings $\phi_i$ together with diagonal sets $D$ would lead to contradictions in $MT_\Delta \overline{A}$! We therefore consider the $Q_i, s$ as "idealized" picture sets of the "physical" sets $\overline{M}_i, \overline{s}$.

The physicist prefers to use a very restricted language which can only be understood by insiders, thus risking misunderstanding. For the relation (cor) of $MT_\Theta$ to the reality, we will also say that the elements of $\overline{M}_i$ and $\overline{s}$ are "physical entities." The problem of knowing whether the elements of $\overline{M}_i$ and the relations $\overline{s}$ can indeed be taken as "real" will be dealt with in Chap. 6. And we will see that indeed this can be done under certain circumstances. But physicists often also designate the elements of $Q_i, s$ as physical entities although they are only "imprecise" pictures of the reality.

How can we describe that the $\phi_i$ are only "imprecise" mappings $\overline{M}_i \to Q_i$? At first we postulate that

the $\phi_i$ are injective. $\hspace{4cm} (3.3.2)$

The postulate (3.3.2) says that the elements of the $Q_i$ can distinguish between different elements of the $\overline{M}_i$, i.e., that two elements of $\overline{M}_i$ will also have different pictures.

Since the $\overline{M}_i$ are finite sets the relation (3.3.2) says practically nothing about the $\phi_i$, since if (3.3.2) was not valid, then we could replace $Q_i$ by a set $Q_i \times R_i$, where the $R_i$ have no influence on the $s$. Thus (3.3.2) does not say anything about the reality; it is only a fixing of the mathematical description.

Instead of the surjectivity, we postulate the weaker relation

$$(\phi_i \overline{M}_i)_{U_i} = Q_i. \tag{3.3.3}$$

The postulate (3.3.3) says only that there are "mathematically" enough elements in $\overline{M}_i$, but not that all elements of $\overline{M}_i$ correspond to "real" objects. But we presume that $\overline{M}_i$ contains only elements that refer to "real" objects. This question will be treated in Chap. 6, where we will consider how to describe what is "real" and not a fairy tale.

We want to express (independently of the $\phi_i$) that a set (also infinite) $X$ (the $Q_i$ or the $s$) is an *idealization of a finite set*. Therefore, we introduce as an additional condition for the definition of an inaccuracy set $U$ for $X$

$$(\exists y)\left[\, y \in X \,\wedge\, y \text{ is finite} \,\wedge\, (y)_U = X \,\right]. \tag{3.3.4}$$

## 3.3 Idealization Process

Now we can formulate the axiom $P_\Delta$:

$$(\exists \phi_1)(\exists \phi_2) \cdots$$
$$[\phi_i : \overline{M}_i \to Q_i \text{ are injective mappings with } (\phi_i \overline{M}_i)_{U_i} = Q_i$$
$$\wedge\ \phi \overline{s} \subset (s)_{U_s}$$
$$\wedge\ \phi \overline{s}' \subset (s')_{U_s} ]. \tag{3.3.5}$$

Here $\overline{s}'$ is the complement of $\overline{s}$ in $\overline{M}_{i_1} \times \cdots \times \overline{\mathbb{R}} \times \cdots$ and $s'$ is the complement of $s$ in $Q_{i_1} \times \cdots \times \mathbb{R} \times \cdots$, and

$$(s)_{U_s} = \{y \mid \exists z\ (z \in s \wedge (y,z) \in U_s)\},$$

$$(s')_{U_s} = \{y \mid \exists z\ (z \in s' \wedge (y,z) \in U_s)\}.$$

For such a test $\Delta$, let us look at the following.

*Example A*

We formulate by $\Delta$ that $\overline{M}, \overline{s}$ is "approximately" a two-dimensional Euclidean geometry:

In $MT_\Theta$ we define the following sets:

$$Q = \mathbb{R} \times \mathbb{R}$$

and $s$ as the set

$$s \subset Q \times Q \times \mathbb{R}$$

of all $(q_1, q_2, d)$ with

$$d(\alpha_1, \alpha_2; \beta_1, \beta_2) = \sqrt{(\alpha_1 - \beta_1)^2 + (\alpha_2 - \beta_2)^2},$$

where $q_1 = (\alpha_1, \beta_1)$ and $q_2 = (\alpha_2, \beta_2)$ with real numbers $\alpha_i, \beta_i$.

We define the following "inaccuracy" set for $Q$:

$$U \subset Q \times Q\ ,\ U_{\varepsilon r} = \left\{ (q_1, q_2) \mid d(q_1, q_2) < \varepsilon \right\} \cup Q_r \times Q_r, \tag{3.3.6}$$

where $Q_r = \left\{ q \mid q \in Q \text{ and } d(q, 0) > r \right\}$; 0 is the point $(0, 0)$ in $\mathbb{R} \times \mathbb{R}$.

For $\mathbb{R}_+$ we introduce the inaccuracy set

$$U_{\delta\varrho} = \left\{(\alpha_1, \alpha_2) \mid |\alpha_1, \alpha_2| < \delta\right\} \cup \mathbb{R}_+^\rho \times \mathbb{R}_+^\rho, \tag{3.3.7}$$

where $\mathbb{R}_+^\rho = \{\alpha \mid \alpha \in \mathbb{R}_+ \wedge \alpha > \rho\}$. We take $\delta < \varepsilon$ and $\varrho > 2r\pi$.

These inaccuracy sets generate in a canonical way also an inaccuracy set $U_s$ in $Q \times Q \times \mathbb{R}_+$. On the basis of the axiom (3.3.5) we have

$(\exists \phi)$
$[\phi : \overline{M} \to Q$ is an injective mapping with $(\phi \overline{M})_U = Q$
$\wedge \ \phi \overline{s} \subset (s)_{U_s}$
$\wedge \ \phi \overline{s}' \subset (s')_{U_s} \ ]. \tag{3.3.8}$

If we want to give particular values for the $\varepsilon, r$ we must take different application domains, one for the round table and one for the surface of the earth:

- For the round table we can choose, e.g., $\varepsilon = 0.1$ mm and $r$ essentially greater than the radius of the round table, e.g., ten times the radius of the table.
- For the surface of the earth we can choose, e.g., $\varepsilon = 10$ cm and $r = 10$ km. But we can also choose other values. If we choose a bigger $r$, we must also choose a bigger $\varepsilon$.

### 3.3.2 Enrichment of $MT_\Delta$ by $\overline{A}$

The introduction of $\overline{A}$ to $MT_\Delta$ is the *same* as the introduction of $\overline{A}$ to $MT_\Theta$ (see Sect. 3.2.3). The only difference is the axiom (3.3.5) and that therefore the addition of $\overline{A}$ can lead to contradictions.

But there arises a major difficulty if we have in some regions of the $Q_i$ "large" inaccuracies as, e.g., $Q_r \times Q_r$ in (3.3.6). We have not introduced any conditions stating that the "real" part (described by $\overline{A}$) does not contain such objects, the picture of which in $Q_i$ is situated in these "large" inaccuracy regions of $Q_i$. Therefore, we have to distinguish two cases for the inaccuracy sets $U$.

The first case is that there are no regions of "large" inaccuracies. We then have to add $\overline{A}$ to $MT_\Delta$ (as in Sect. 3.2.3) and must see whether there is a contradiction to the axiom (3.2.5). If there is no contradiction to the axiom, the test of the theory is positive. But what is necessary for this test?

The text $\overline{A}$ is given by relations of the form (3.2.9), (3.2.10), and (3.2.11). If we write $\phi \overline{a} = a$, these relations take the form of the following relations:

$a \in Q,$
$(a_1, a_2, \cdots, J) \cap \phi \overline{s} \neq \emptyset, \tag{3.3.9}$
$(a_1, a_2, \cdots, J) \subset \phi \overline{s}'.$

We have no contradiction to the axiom (3.3.5) if the relations of the form

$a \in Q$,

$(a_1, a_2, \cdots, J) \cap (s)_{U_s} \neq \emptyset$, \hfill (3.3.10)

$(a_1, a_2, \cdots, J) \subset (s')_{U_s}$

do not lead to contradictions in $MT$ (i.e., without $\Theta$!). Here we have to take the $a_i$ as new constants to $MT$. For (3.3.10) we write briefly $A$ and $MTA$ for the mathematical theory $MT$ with the addition of the text (3.3.10). The test is positive if $MTA$ is without contradiction.

(The relation $A$ was introduced in [1] Chap. 6 by "the principles of imprecise mapping" formulated intuitively.)

We conclude that in the first case, it suffices for a test to see whether $MTA$ is without contradiction.

The second case is that there are regions of "large" inaccuracies as, e.g., in our example A. Thus it can be that there exists a mapping $\phi$ which fulfills the axiom (3.3.5) and which maps all the $\bar{a}$ of $\overline{A}$ in regions of "large" inaccuracies, so that there is no contradiction between $\overline{A}$ and (3.3.5). But for such a theory we do not pretend that the theory describes the reality in a global region, but only in a subregion of the application domain (which can also be in various subregions!) where it is possible to use only small inaccuracy sets. How can we formulate this in the language of $MT_\Delta$?

We choose one (or some) of the tuples $(\bar{a}_1, \bar{a}_2, \ldots) \in \overline{M}_1 \times \overline{M}_2 \times \cdots$. Such a tuple is not only a mathematical symbol, but it is also a part of $\overline{A}$, i.e., a representation of established real objects. We now change the axiom (3.3.5) by adding between the brackets of (3.3.5) the condition for $\phi$

$$\phi(\bar{a}_1 \times \bar{a}_2, \cdots) \in \widehat{Q} \subset Q_1 \times Q_2 \cdots, \hfill (3.3.11)$$

where $\widehat{Q}$ is a part where the inaccuracies are small. Sometimes we choose a particular element $\widehat{q} \in \widehat{Q}$ and write instead of (3.3.11)

$$\phi(\bar{a}_1 \times \bar{a}_2, \cdots) = \widehat{q}. \hfill (3.3.12)$$

If we do not find by enrichments of $\overline{A}$ (by the addition of new experiments) a contradiction with (3.3.5) and (3.3.11) (respectively (3.3.5) and (3.3.12)) we say that $MT_\Delta$ describes with small inaccuracies a surrounding of $\widehat{Q}$ (respectively $\widehat{q}$).

It can be that we can use the theory for more than one $\widehat{Q}$ (or $\widehat{q}$). We will demonstrate this below.

*Example A*

In this case we record - marked spots - and the results of measurements of the - distances between these marked spots -.

We presuppose the knowledge of "how to measure a distance," i.e., the knowledge of a pre-theory of measurements of distances, a pre-theory of the use of such measuring records, but without the geometry of the surfaces. These pre-theories show that there are no precise measurements. We say that every measurement has a so-called "error of measurement" describing the imprecision of the measurement. The result of a measurement of the distance of two marked spots $\bar{a}, \bar{b}$ may be given by the real number $\alpha$. We then have to correct this by saying that the value is not exactly $\alpha$, but can be any number between $\alpha - \varepsilon$ and $\alpha + \varepsilon$, i.e., a number of an interval $J$.

Now we want to describe the use of our theory $MT_\Delta$ for the description of the real application domain of marked spots on the earth, and their measured distances.

We see by (3.3.6) that in this case we have an example of the "second case" of Sect. 3.3.2. The region of "large" inaccuracies is given by $Q_r$, and the point $(0,0) \in Q$ is a "center" of the region $Q'_r$ of "small" inaccuracies. We choose one $\bar{a}$ of the marked spots appearing in $\bar{A}$. The selected $\bar{a}$ may be $\bar{a}_0$ ($\bar{a}_0$ can be, e.g., a particular marked spot at Greenwich in London). We add in (3.3.8) the condition

$$\phi \bar{a}_0 = (0,0).$$

Thus we get the axiom

$$\begin{aligned}&(\exists \phi)\\ &[\phi : \overline{M} \to Q \text{ is an injective mapping with } (\phi \overline{M})_U = Q\\ &\wedge\ \phi \bar{a}_0 = (0,0)\\ &\wedge\ \phi \bar{s} \subset (s)_{U_s}\\ &\wedge\ \phi \bar{s}' \subset (s')_{U_s}\ ]. \end{aligned} \qquad (3.3.13)$$

A text $\bar{A}$ consists of relations of the form

$$\bar{a} \in \overline{M} \quad \text{and} \quad (\bar{a}_1, \bar{a}_2, J_{12}) \cap \bar{s} \neq \emptyset$$

which lead, for a $\phi$ satisfying (3.3.13), to relations of the form

$$\phi \bar{a} \in Q \quad \text{and} \quad (\phi \bar{a}_1, \phi \bar{a}_2, J_{12}) \cap (s)_{U_s} \neq \emptyset. \qquad (3.3.14)$$

(Relations for $\bar{s}'$ and $s'$ are not interesting for us because they are mathematically irrelevant.) If we write $\phi \bar{a} = a$, we get relations of the form

$$\bar{a}_0 = (0,0) \ , \ \bar{a} \in Q \quad \text{and} \quad (\phi\bar{a}_1, \phi\bar{a}_2, J_{12}) \cap (s)_{U_s} \neq \emptyset. \tag{3.3.15}$$

Equation (3.3.15) is only interesting for such an $a$ with $a \in Q'_r$, i.e., for the surrounding of $a_0 = (0,0)$.

In this example, a mapping $\phi$ satisfying (3.3.13) has a very graphic representation: the application $\phi$ generates a map of the earth onto a sheet of graph paper. On this map, the surrounding of $\bar{a}_0$ (Greenwich in London) represents the distances between various marked spots very well, but very badly the marked spots situated very far away from $\bar{a}_0$.

It is well known that we can have many such maps by choosing different $\bar{a}_0$.

Because the surroundings of the various $\bar{a}_0$ are "well" represented by the various maps, we often only print these surroundings on the graph paper. This leads us to the problem of the following section (i.e., 3.3.3).

Our example of the marked spots on the surface of the earth appears artificial. However, there is a very interesting example which is similar to the example given above:

In the application domain $A_p$ we take space–time marked spots, which are usually called "events." In $MT_\Theta$ we introduce as a constant the set $\overline{M}$ with the interpretation

'$\bar{a} \in \overline{M}$' *corresponds to* '$\bar{a}$ *is an event*'.

We want to give for the Galileo–Newton theory of space–time (briefly *GN* theory) the form $MT_\Delta$. In $MT_\Theta$ we introduce two relations: $\bar{s}_1 \subset \overline{M} \times \overline{M} \times \mathbb{R}$ and $\bar{s}_2 \subset \overline{M} \times \overline{M} \times \mathbb{R}$ with the interpretations

- '$(\bar{a}_1, \bar{a}_2, \alpha) \in \bar{s}_1$' *corresponds to* '$\alpha$ *is a possible value for the spatial distance of the events* $\bar{a}_1$ *and* $\bar{a}_2$';
- '$(\bar{a}_1, \bar{a}_2, \beta) \in \bar{s}_2$' *corresponds to* '$\beta$ *is a possible value for the time distance of the events* $\bar{a}_1$ *and* $\bar{a}_2$'.

We assume that we know by pre-theories how to measure these relations.

For $MT_\Delta$ we add the following terms:

$Q = \mathbb{R} \times \mathbb{R} \times \mathbb{R} \times \mathbb{R}$ as picture term for $\overline{M}$,

$s_1 = Q \times Q \times \mathbb{R}$ as picture term for $\bar{s}_1$,

$s_2 = Q \times Q \times \mathbb{R}$ as picture term for $\bar{s}_2$,

where

$$s_1 = \big\{((\alpha_1, \alpha_2, \alpha_3, \alpha_4), (\beta_1, \beta_2, \beta_3, \beta_4), d) \ \big|$$
$$d^2 = (\alpha_1 - \beta_1)^2 + (\alpha_2 - \beta_2)^2 + (\alpha_3 - \beta_3)^2\big\}$$

and

$$s_2 = \{((\alpha_1,\alpha_2,\alpha_3,\alpha_4),(\beta_1,\beta_2,\beta_3,\beta_4),t) \mid t^2 = \beta_4 - \alpha_4\}.$$

The choice of the inaccuracy set $U$ for $Q$ is very similar to (3.3.6). We have to choose not only $d(q_1,q_2) < \varepsilon$ but also $t < \tau$. Instead of $Q_r$ in (3.3.6) we take

$$Q_{rT} = \{q \mid q \in Q \text{ and } d(q,0) > r\} \cup \{q \mid q \in Q \text{ and } t(q,0) > T\}.$$

If we take for the mapping $\phi : \overline{M} \to Q$ the condition that the event $\overline{a} = \{$a certain time at the center of mass of our planetary system$\}$ is mapped onto $(0,0,0,0) \subset Q$, then we get the space–time system of Galileo, Kepler, and Newton as a pre-theory for the description of the motion of the planets. This system is nothing other than a particular "map" of the surrounding of $\overline{a}$. For other regions in the universe it will be better to use other "maps," and for the surrounding of a "black hole" we will have no "map."

### 3.3.3 Fundamental Domain of a *PT*

We have seen in Sect. 3.3.2 that the fact of having a useful theory (i.e., that $MT_\Delta \overline{A}$ is without contradiction) depends on the choice of the inaccuracy sets $U$ for $\Delta$. We have also seen that it is necessary to distinguish between two cases: Either there are or there are *not* "large" inaccuracies. If there are no large inaccuracies, we say that the theory can be applied as a "good" description of the application domain $A_p$. But if there are large inaccuracies, we say that the theory says practically nothing about the structure of reality in such regions where the inaccuracies are large.

Thus, it is more useful to apply a theory only on such a part of the application domain $A_p$ where we can use inaccuracy sets $U$ with "small' inaccuracies. In such a region, the theory essentially says something about the structure of reality and will be useful for technical applications. We call such a region (a part of the application domain $A_p$) the *fundamental domain* $G$. (For the theories where we could use in the total application domain $A_p$ "small" inaccuracies we have $G \equiv A_p$.)

How can we select from $A_p$ a fundamental domain $G$?

The starting point is not a part of the application domain $A_p$ but a part of $Q_1 \times Q_2 \times \cdots$ On the basis of the inaccuracy sets $U$ we choose a part $F \subset Q_1 \times Q_2 \times \cdots$ for which $U$ describes only "small" inaccuracies. Now, how do we go from $F$ to the fundamental domain $G$? At first we must select a particular mapping $\phi$ by a procedure already described in Sect. 3.3.2. We choose in the application domain $A_p$ particular objects $\overline{a}_0 \in \overline{M}_i$ and postulate that the $\phi(\overline{a}_0)$ determines an element in the "middle" of $F$. Then $\overline{F} = \phi^{-1} F$ is a part of $\overline{M}_1 \times \overline{M}_2 \times \ldots$. An $(\overline{a}_1, \overline{a}_2, \cdots)$ of the application domain $A_p$ lies by (cor) in $\overline{F}$. If we can show by the theory and by experiments (that is by $\overline{A}$) that (cor) gives $(\overline{a}_1, \overline{a}_2, \cdots) \in \overline{F}$, then we say that $(\overline{a}_1, \overline{a}_2, \cdots)$ lies in the fundamental domain $G$ as a part of the application domain $A_p$, i.e., $G(\text{cor})\overline{F}$.

*Example A*

If we consider as an application domain $A_p$ the marked spots on the surface of the earth, from Sect. 3.3.2 we see immediately that a fundamental domain $G$ is given by the surrounding of a particular marked spot (e.g., a marked spot at Greenwich) from which we get by $(\text{cor})\phi$ a "map" with small inaccuracies relative to the distances.

We also see that it is possible to have more than one fundamental domain $G$. Therefore, it is necessary for the application of a theory to give the selected fundamental domain $G$. If we apply, e.g., the $GN$ space–time theory, we have to give the fundamental domain $G$ as described in Sect. 3.3.2, e.g., the space–time region in our planetary system. The $GN$ space–time theory says nothing about the global structure of space–time, and Newton had already suspected that his theory could be wrong for the global structure of the universe.

*Example A*

In the case of the surface of a round table, we have as the fundamental domain $G$ the total surface, i.e., the total application domain $A_p$. The problem with this example is the following: Which part of $\overline{M}$ refers to "real" marked spots? We will return to this problem in Chap. 6.

There will be later additional aspects by which we will restrict the application domain $A_p$ to the fundamental domain $G$ (see Sect. 4.8).

# 4
# Species of Structures and Axiomatic Basis of a $PT$

In the preceding chapter we described the general form of a mathematical theory in the context of a physical theory. The main task was to clarify the connection between physical realities and the abstract mathematical theory. Nevertheless, this explanation was not sufficient to apprehend the structure of the mathematical theory itself; it was especially not sufficient to see how $MT_\Delta$ gives the physically decisive conditions to $MT_\Theta$. This problem was hidden behind the selection of the picture terms $Q_i$ and $s_\nu$ in $MT$; for example, as picture terms $Q = \mathbb{R} \times \mathbb{R}$ for the idealized marked spots, and as relation $s$

$$d(\alpha_1, \alpha_2; \beta_1, \beta_2) = \sqrt{(\alpha_1 - \beta_1)^2 + (\alpha_2 - \beta_2)^2}$$

for the idealized distance. In this case we know that the terms $Q, d$ describe a two-dimensional Euclidean geometry, but we only know this since we have learned that the terms $Q, d$ constitute the analytic form of this geometry.

This example also shows that the particular values $(\alpha_1, \alpha_2)$ for an idealized marked spot have no influence on the physically interpreted part $MT_\Theta$. For instance, if we make the transformation

$$\alpha'_1 = \alpha_1 \cos\varphi + \alpha_2 \sin\varphi,$$
$$\alpha'_2 = -\alpha_1 \sin\varphi + \alpha_2 \cos\varphi,$$

from $Q$ to $Q'$, this has no influence on the "real" marked spots and distances described by $MT_\Theta$.

Thus we get the question: "What are the essential parts of the selection of the picture terms $Q_i$ and $s_\nu$ relative to $MT_\Theta$?" In many cases, one speaks in $MT$ of what we have called in Sect. 1.3 fairy tales. There is also another question: "By what does one often try to find intuitively the picture terms $Q_i$ and $s_\nu$?" For these questions we will try to find a suitable form for the $MT$ in $MT_\Theta$. The basis of this purpose is the mathematical concept of a species of structures.

## 4.1 Mathematical Structures

At first we recall the well-known definition of a species of structures:

Let $MT$ be a theory which is stronger than the set theory, let $x_1, \ldots, x_n, s$, be letters which are distinct from the constants of $MT$, and let $A_1, \ldots, A_m$ be terms in $MT$ (in which none of the letters $x_1, \ldots, x_n, s$ appears). Let $S$ be an *echelon construction scheme* (see the end of Sect. 2.4) on $n + m$ terms. The relation

$$T(x_1, \ldots, x_n, s) \; : \; s \in S(x_1, \ldots, x_n, A_1, \ldots, A_m)$$

is called a *typification* of the letter $s$. The term $s$ is called a *structure term*. A relation $P(x_1, \ldots, x_n, s)$ is said to be *transportable* with respect to the typification $T(x_1, \ldots, x_n, s)$ if bijective mappings lead to equivalent relations, in other words if $MT$ contains the following theorem:

From

"$T(x_1, \ldots, x_n, s) \wedge (f_1$ is a bijective mapping of $x_1$ onto $y_1)$

$\wedge \ldots$

$\wedge (f_n$ is a bijective mapping of $x_n$ onto $y_n)$"

follows the relation

$$P(x_1, \ldots, x_n, s) \Leftrightarrow P(y_1, \ldots, y_n, s'),$$

where

$$s' = \langle f_1, \ldots, f_n, \mathrm{Id}_1, \ldots, \mathrm{Id}_m \rangle^S (s)$$

and $\mathrm{Id}_i$ ($1 \leq i \leq m$) denotes the identity mapping of $A_i$ onto itself. In the sequel, the relation $P$ will be called an *axiomatic relation* in order to underline the fact that this relation will be postulated later as an axiom.

We take as text $\Sigma$ the following groups of symbols: the letters $x_1, \ldots, x_n, s$, the typification $T(x_1, \ldots, x_n, s)$, and the axiomatic relation $P(x_1, \ldots, x_n, s)$. This text $\Sigma$ is called a *species of structures*. The letters $x_1, \ldots, x_n$ are called the *principal base sets*, the letters $A_1, \ldots, A_m$ the *auxiliary base sets*, and the letter $s$ the *structure term of the species $\Sigma$*.

If one adds the relation "$T \wedge P$" as an axiom to $MT$, then one obtains a theory $MT_\Sigma$ which is stronger than $MT$. The constants of $MT_\Sigma$ are therefore the constants of $MT$ and the letters which appear in $T$ or in $P$, i.e., the letters $x_1, \ldots, x_n, s$. The theory $MT_\Sigma$ is called the *theory of the species of structures $\Sigma$*.

To summarize: for physical theories we can specify the form of a species of structures. Every mathematical theory $MT$, as a part of a physical theory, will contain the set theory. (It contains therefore also the theories of real and complex numbers.) The additional element is the structure, which is a text $\Sigma$, given by

1. A certain number of *principal base sets* and *auxiliary base sets* $x := \langle x_1, \ldots, x_n \rangle$, $A := \langle A_1, \ldots, A_m \rangle$.
2. A finite number of *structure terms* $s := \langle s_1, \ldots, s_p \rangle$
3. with the *typification*

$$T(x,s) \;:\; \bigwedge_{\nu=1}^{p} s_\nu \subset S_\nu(x, A),$$

with $p$ echelon construction schemes $S_\nu$.

4. The *axiomatic relation*

$$P(x,s) \;:\; \bigwedge_{\mu=1}^{q} \alpha_\mu(x, A, s)$$

composed of axiomatic relations $\alpha_\mu$.

For this species of structures $\Sigma$ we write

$$\Sigma(x,s) \equiv \Sigma\,(\,T(x,s) \,\wedge\, P(x,s)\,),$$

and for the mathematical theory $MT$ endowed with this species of structures $\Sigma(x,s)$ we write

$$MT_{\Sigma(x,s)}.$$

It is not difficult to see that we can write this $\Sigma(x,s)$ also in the general mathematical form of a species of structures given above. We only have to introduce instead of the $s_\nu$ one $s$ with the typification

$$s \in \mathcal{P}(\mathcal{P}S_1(x, A) \times \mathcal{P}S_2(x, A) \times \cdots \times \mathcal{P}S_p(x, A)),$$

i.e.,

$$s \subset \mathcal{P}S_1(x, A) \times \mathcal{P}S_2(x, A) \times \cdots \times \mathcal{P}S_p(x, A),$$

i.e.,

$$s := \langle s_1, \ldots, s_p \rangle \quad \text{with} \quad s_\nu \in \mathcal{P}S_\nu(x, A),$$

i.e.,

$$s_\nu \subset S_\nu(x, A).$$

## 4 Species of Structures and Axiomatic Basis of a *PT*

The reason for which we have introduced the concept of species of structures by the text above is easy to see: $\Theta$ of Sect. 3.2.2 and $\Delta$ of Sect. 3.3.1 are such species of structures.

For $\Theta$ we have

1. the principal base sets $\overline{M}_i$ and the auxiliary base set $\overline{\mathbb{R}}$,
2. the structure term $\overline{s}$,
3. with the typifications $\overline{s} \subset \overline{M}_i$ or $\overline{s} \subset \overline{M}_{i_1} \times \overline{M}_{i_2} \times \cdots \times \overline{\mathbb{R}} \times \cdots$,
4. *without* an axiomatic relation $P$.

For $\Delta$ we have

1. the principal base sets $\overline{M}_i$ as for $\Theta$ and the auxiliary base sets $\overline{\mathbb{R}}$, $\mathbb{R}$, $Q_i$, various $s$, the $U_i$ and $U_s$,
2. the structure terms $\overline{s}$ as for $\Theta$,
3. with the typifications $\overline{s} \subset \overline{M}_i$ or $\overline{s} \subset \overline{M}_{i_1} \times \overline{M}_{i_2} \times \cdots \times \overline{\mathbb{R}} \times \cdots$ as for $\Theta$,
4. *with* the axiomatic relation $P_\Delta$ given in (3.3.5).

Now we want to introduce a relation between two species of structures which is often used, especially in the mathematical parts of a physical theory.

We consider two species of structures $\Sigma_1$ and $\Sigma_2$ with the same principal base sets, the same auxiliary base sets, the same typification, but with different axiomatic relations $P_1$ and $P_2$. If $P_1$ is a theorem in $MT_{\Sigma_2}$, then we say that the species of structures $\Sigma_2$ is *richer* (or *finer*) than $\Sigma_1$, or that $\Sigma_1$ is *poorer* (or *coarser*) than $\Sigma_2$. If $P_1$ is a theorem in $MT_{\Sigma_2}$ and $P_2$ is a theorem in $MT_{\Sigma_1}$, then we say that $\Sigma_1$ and $\Sigma_2$ are *equally rich*, and sometimes also simply "equal."

We want to describe a simple example of a species of structures $\Sigma$: the species of "lattice" structures. We take one principal base set $x$ (with no auxiliary base set). The order relation "<" is defined by a structure term $s$ with the typification

$$T(x,s) \;:\; s \subset x \times x,$$

that is, the relation $(y_1, y_2) \in s$ is noted $y_1 < y_2$. In the axiomatic relation $P(x,s)$, one groups all the axioms for the order relation "<" that one writes "for a lattice $x$." The transition from a set "only" ordered to a lattice, and then to a distributive lattice is an example of a transition from a species of structures to a species of structures increasingly richer.

A theory $MT$ can already contain a species of structures $\Sigma$. This means, a term (a set) $V$ of $MT$ is said to be a structure of species $\Sigma$ on the base sets $E_1, \ldots, E_n$ which are also sets in $MT$ if the relation

$$\text{``}T(E_1, \ldots, E_n, V) \;\wedge\; P(E_1, \ldots, E_n, V)\text{''}$$

is a theorem in $MT$. We will often say that "the base sets $E_1, \ldots, E_n$ are endowed with the structure $V$ of species $\Sigma$." Thus, $V$ is an element of $S(E_1, \ldots, E_n, A_1, \ldots, A_m)$ which satisfies the relation $P(E_1, \ldots, E_n, V)$. Let $W$ be the subset of all elements $W$ of $S(E_1, \ldots, E_n, A_1, \ldots, A_m)$ which satisfies the relation $P(E_1, \ldots, E_n, V)$. Then it follows that $W$ is the set of all structures of the species $\Sigma$ on the base sets $E_1, \ldots, E_n$. For each theorem $B(x_1, \ldots, x_n, s)$ of $MT_\Sigma$ it follows for any element $V$ of $W$ that $B(E_1, \ldots, E_n, V)$ is also a theorem in $MT$.

If $f_1, \ldots, f_n$ are bijective mappings in $MT$ of the base sets $E_1, \ldots, E_n$ of the structure $V$ onto other sets $E'_1, \ldots, E'_n$, then one has (since the axiomatic relation $P$ is transportable) that the set $V'$, defined by

$$V' = \langle f_1, \ldots, f_n, \mathrm{Id}_1, \ldots, \mathrm{Id}_m \rangle^S (V),$$

is a structure of the same species $\Sigma$ on the base sets $E'_1, \ldots, E'_n$.

In $MT$, let $V$, $V'$ be structures of species $\Sigma$ on the base sets $E_1, \ldots, E_n$ and $E'_1, \ldots, E'_n$ respectively, and let $f_1, \ldots, f_n$ be bijective mappings of the base sets $E_1, \ldots, E_n$ onto the base sets $E'_1, \ldots, E'_n$ such that

$$V' = \langle f_1, \ldots, f_n, \mathrm{Id}_1, \ldots, \mathrm{Id}_m \rangle^S (V),$$

then $(f_1, \ldots, f_n)$ is said to be an *isomorphism* of the base sets $E_1, \ldots, E_n$ onto the base sets $E'_1, \ldots, E'_n$, endowed with structures $V$ and $V'$ respectively. In this case, the structures $V$ and $V'$ are said to be *isomorphic*, and the bijective mapping $f_1, \ldots, f_n$ is an isomorphism of $V$ and $V'$.

If every two structures of the species $\Sigma$ are isomorphic, then the *species of structures $\Sigma$* is said to be *univalent* or *categorical*; otherwise *polyvalent*.

## 4.2 Deduction of Structures

In the preceding paragraph we introduced the notion of species of structures, and we saw that structures of such a species can exist in $MT$. Let us examine now particular structures of species $\Sigma'$ in $MT_\Sigma$.

Let us consider two species of structures $\Sigma$, $\Sigma'$ in $MT$ characterized as follows:

- for the species $\Sigma$ : $n$ principal base sets $x_1, \ldots, x_n$, $m$ auxiliary base sets $A_1, \ldots, A_m$, a structure term $s$ with the typification

$$T(x_1, \ldots, x_n, s) \ : \ s \in S(x_1, \ldots, x_n, A_1, \ldots, A_m),$$

- for the species $\Sigma'$ : $r$ principal base sets $y_1, \ldots, y_r$, $p$ auxiliary base sets $B_1, \ldots, B_p$, a structure term $t$ with the typification

$$T(y_1, \ldots, y_r, t) \ : \ t \in S'(y_1, \ldots, y_r, B_1, \ldots, B_p);$$

- nothing is presupposed about the axiomatic relations $P$ of $\Sigma$ and $P'$ of $\Sigma'$!

Then a system of $r+1$ terms $E_1, \ldots, E_r, V$ such that

1. $V$ is a structure of species $\Sigma'$ on $E_1, \ldots, E_r$ in $MT_\Sigma$,
2. the terms $E_1, \ldots, E_r, V$ are intrinsic in $MT_\Sigma$

is called a *procedure of deduction of a structure of species $\Sigma'$ from a structure of species $\Sigma$*.

A term $W(x_1, \ldots, x_n, s)$ is said to be *intrinsic* if it is an element of an echelon set on the base sets $x_1, \ldots, x_n, A_1 \ldots, A_m$ and if, by bijective mappings $f_1, \ldots, f_n$ of the base sets $x_1, \ldots, x_n$ onto the base sets $x'_1, \ldots, x'_n$, the canonical image of $W(x_1, \ldots, x_n, s)$ becomes $W(x'_1, \ldots, x'_n, s')$, where $s'$ is the canonical image of $s$.

The procedure of deduction of a structure of species $\Sigma'$ from a structure of species $\Sigma$ is then determined by the terms

$$E_j = E_j(x_1, \ldots, x_n, s) \ (1 \leq j \leq r), \quad V = V(x_1, \ldots, x_n, s).$$

These terms indicate the methods whereby the terms are deduced. It will often be a mathematical problem to find such a procedure of deduction. In [8], e.g., we deduced the Hilbert's species of structures of space $\Sigma'$ from a species of structures $\Sigma$ "nearer to physics."

A reason why we have only taken intrinsic terms for the procedure of deduction of structures is the following:

We start now from two structures $s, s'$ of the same species $\Sigma$. Let $x_1, \ldots, x_n$ be the base sets for $s$, let $x'_1, \ldots, x'_n$ be the base sets for $s'$, and let $(f_1, \ldots, f_n)$ be an isomorphism of $s$ onto $s'$. The system of $r+1$ terms $E_i, V$ is a procedure of deduction of a structure $V$ of another species $\Sigma'$.

Let, with the echelon construction scheme $T_i$, the type of $E_i$ be described by

$$E_i \in \mathcal{P}(T_i(x_1, \ldots, x_n, A_1, \ldots, A_m)),$$

i.e., the terms $E_i$ are subsets of the $T_i(\cdots)$. Since the $E_i, V$ are intrinsic terms, the mappings

$$g_i = \langle f_1, \ldots, f_n, \mathrm{Id}_1, \ldots, \mathrm{Id}_m \rangle^{T_i} \quad (1 \leq i \leq r)$$

of $E_i = E_i(x_1, \ldots, x_n, s)$ onto $E'_i = E'_i(x'_1, \ldots, x'_n, s')$ form an isomorphism of $V = V(x_1, \ldots, x_n, s)$ onto $V' = V'(x'_1, \ldots, x'_n, s')$.

This applies naturally to the following case: In a theory $MT$, let $W$ be a structure of species $\Sigma$ on the base sets $F_1, \ldots, F_n$, let $W'$ be a structure of the same species $\Sigma$ on the base sets $F'_1, \ldots, F'_n$, and let $f_1, \ldots, f_n$ (mappings of the base sets $F_1, \ldots, F_n$ onto the base sets $F'_1, \ldots, F'_n$) be an isomorphism of $W$ onto $W'$. If the system of $r+1$ terms $E_i, V$ is a procedure of deduction of structures of the species $\Sigma'$, then with

$$E_i = E_i(F_1, \ldots, F_n, W), \quad V = V(F_1, \ldots, F_n, W),$$

$$E'_i = E_i(F'_1, \ldots, F'_n, W'), \quad V' = V(F'_1, \ldots, F'_n, W'),$$

and with $E_i \subset T_i(F_1, \ldots, F_n, A_1, \ldots, A_m)$ we get that

$$g_i = \langle f_1, \ldots, f_n, \text{Id}_1, \ldots, \text{Id}_m \rangle^{T_i}$$

is an isomorphism of the structures $V, V'$ of species $\Sigma'$.

We want to define the equivalence of two species of structures $\Sigma$ and $\Sigma'$. At first $\Sigma$ and $\Sigma'$ will have the same base sets $x_1, \ldots, x_n$: $\Sigma(x_1, \ldots, x_n, s)$ and $\Sigma'(x_1, \ldots, x_n, t)$.

There may be in $MT_\Sigma$ a procedure of deduction $V$ of species $\Sigma'$

$$E_i = x_i, \quad V = V(x_1, \ldots, x_n, s),$$

and vice versa there may be in $MT'_\Sigma$ a procedure of deduction $W$ of species $\Sigma$

$$E'_i = x_i, \quad W = W(x_1, \ldots, x_n, t).$$

We say that the species of structures $\Sigma$ and $\Sigma'$ are *equivalent* relative to the procedures $V(x_1, \ldots, x_n, s)$ and $W(x_1, \ldots, x_n, t)$ if $V = t$ and $W = s$, i.e., if

$$V(x_1, \ldots, x_n, W(x_1, \ldots, x_n, t)) = t, \quad W(x_1, \ldots, x_n, V(x_1, \ldots, x_n, t)) = s.$$

For every theorem $A(x_1, \ldots, x_n, s)$ in $MT_\Sigma$, the relation $A(x_1, \ldots, x_n, W)$ is a theorem in $MT_{\Sigma'}$ and, conversely, for every theorem $B(x_1, \ldots, x_n, t)$ in $MT_{\Sigma'}$ the relation $B(x_1, \ldots, x_n, V)$ is a theorem in $MT_\Sigma$.

From the observations above on isomorphisms by deduced structures, it follows that in a theory $MT'$, if $S, S'$ are structures of species $\Sigma$ on the base sets $E_1, \ldots, E_n$ and $E'_1, \ldots, E'_n$ respectively, and if $S_0, S'_0$ are structures of species $\Sigma_0$ equivalent to $S$ and $S'$ respectively, then $(f_1, \ldots, f_n)$ is an isomorphism of $S$ onto $S'$ if and only if it is an isomorphism of $S_0$ onto $S'_0$. For this reason, we make no distinction between the theories $MT_\Sigma$ and $MT_{\Sigma'}$ of two equivalent species of structures. We consider these theories as a *single* theory, with a single name, e.g., the theory of species of "topological space" structures.

The same theory $MT$ endowed with species of structures $\Sigma$ and $\Sigma'$ which are equivalent is frequently used. As an example, we briefly refer to the theory of topological spaces: As $MT$ we take the theory of sets. Let $\Sigma$ be the species of structures on the single principal base set $x$. The typification is

$$T(x,s): s \in \mathcal{PP}(x), \quad \text{i.e.,} \quad s \subset \mathcal{P}(x),$$

the structure term $s$ is the set of open sets for which one introduces as an axiomatic relation

$$P(x,s) \; : \; \text{conditions on open sets.}$$

The set $x$ is called "topological space." Let $\Sigma'$ be the species of structures on the *same* single principal base set $x$. The typification is

$$T(x,t): t \in \mathcal{P}(x \times \mathcal{PP}(x)), \quad \text{i.e.,} \quad t \subset x \times \mathcal{PP}(x),$$

the structure term $t$ is the set of all pairs $(y, V(y))$ with $y \in x$ and $V(y)$ is the neighborhood filter of $y$ for which one introduces as an axiomatic relation

$$P(x,t) \; : \; \text{conditions on systems of neighborhoods.}$$

The set $t$ is called "set of neighborhoods." In order to show the equivalence of $\Sigma$ and $\Sigma'$ in this example, one defines

- in $MT_\Sigma$ the set $V(x,s)$ of neighborhoods, i.e., a structure $V$ of species $\Sigma'$;
- in $MT_{\Sigma'}$ the set $W(x,t)$ of open sets, i.e., a structure $W$ of species $\Sigma$.

Then one shows that $W(x, V(x,s)) = s$. In the same way, one shows that $V(x, W(x,t)) = t$, and thus the equivalence of $\Sigma$ and $\Sigma'$.

But in physics, there is a case even more general of the relation between two species of structures $\Sigma$, $\Sigma'$ which is of great importance. We assume that a deduction of a species of structures $\Sigma$ is given in a theory $MT_{\Sigma'}$. Let $\Sigma'$ be the species of structures characterized by the principal base sets $x_1, \ldots, x_n$, the auxiliary base sets $A_1, \ldots, A_m$, the structure term $s \in S(x_1, \ldots, x_n, A_1, \ldots, A_m)$, and the axiomatic relation $P'(x_1, \ldots, x_n, s)$. The deduction of $\Sigma$ from $\Sigma'$ consists in the specification of intrinsic terms

$$E_j = E_j(x_1, \ldots, x_n, s) \quad (1 \leq j \leq r),$$

which are principal base sets of the structure $V = V(x_1, \ldots, x_n, s)$ of species $\Sigma$ in $MT_{\Sigma'}$ (therefore $V(x_1, \ldots, x_n, s)$ is also an intrinsic term). It could be that there exists a species of structures $\Sigma_1$, richer than $\Sigma$, such that the structure $V$ is also a structure of species $\Sigma_1$. How can one be certain that the structure $V$ which has been deduced is "only" a structure of species $\Sigma$ and "not" a structure of richer species?

For this, we introduce the following condition. The structure $V$ of species $\Sigma$ deduced in $MT_{\Sigma'}$ is called a *representation* of $\Sigma$ in $\Sigma'$ if the following condition is satisfied: In $MT_\Sigma$, let $y_1,\ldots,y_n$ be the principal base sets, let $A_1,\ldots,A_m$ be the auxiliary base sets, let $t$ be the structure term, and let $P(y_1,\ldots,y_r,t)$ be the axiomatic relation of $\Sigma$. We can demonstrate the following theorem:

$$(\exists x_1)\cdots(\exists x_n)(\exists s)(\exists f_1)\cdots(\exists f_r)$$
$$\bigl[\,T'(x_1,\ldots,x_n,s)\colon s\in S'(x_1,\ldots,x_n,A_1,\ldots,A_m)$$
$$\wedge\ P'(x_1,\ldots,x_n,s)$$
$$\wedge\ f_i:y_i\to E_i(x_1,\ldots,x_n,s)\ \text{are bijective mappings}$$
$$\text{with}\ \langle f_1,\ldots,f_r,\mathrm{Id}_1,\ldots,\mathrm{Id}_m\rangle^S t = V(x_1,\ldots,x_n,s)\,\bigr]. \qquad (4.2.1)$$

In the notion of representation appears explicitly the deduction

$$E_j = E_j(x_1,\ldots,x_n,s)\ (1\le j\le r),\ V = V(x_1,\ldots,x_n,s)$$

of the species of structures $\Sigma$ in $MT_{\Sigma'}$.

If $V(x_1,\ldots,x_n,s)$ is a representation of $\Sigma$ in $MT_{\Sigma'}$ and $R(y_1,\ldots,y_r,t)$ a transportable relation with which the axiomatic relation of $\Sigma$ can be enriched, and if we can deduce $R(E_1,\ldots,E_r,V)$ as a theorem in $MT_\Sigma$, then we have, with constants $x_1,\ldots,x_n,s$ not appearing in $MT_{\Sigma'}$:

$$\bigl[\,T'(x_1,\ldots,x_n,s)\ \wedge\ P'(x_1,\ldots,x_n,s)\,\bigr]\ \Rightarrow\ R(E_1,\ldots,E_r,V)$$

is a theorem in $MT_\Sigma$. Moreover, with other constants $f_1,\ldots,f_r$ not appearing in $MT_\Sigma$:

$$(\exists x_1)\cdots(\exists x_n)(\exists s)(\exists f_1)\cdots(\exists f_r)$$
$$\bigl[\,T'(x_1,\ldots,x_n,\varepsilon):s\in S'(x_1,\ldots,x_n,A_1,\ldots,A_m)$$
$$\wedge\ P'(x_1,\ldots,x_n,s)$$
$$\wedge\ \neg R(y_1,\ldots,y_r,t)$$
$$\wedge\ f_i:y_i\to E_i(x_1,\ldots,x_n,s)\ \text{are bijective mappings}$$
$$\text{with}\ \langle f_1,\ldots,f_r,\mathrm{Id}_1,\ldots,\mathrm{Id}_m\rangle^S t = V(x_1,\ldots,x_n,s)\,\bigr]$$
$$\Rightarrow\ \neg R(E_1,\ldots,E_r,V)$$

is a theorem in $MT_\Sigma$. Consequently, the relation

$$(\exists x_1)\cdots(\exists x_n)(\exists s)(\exists f_1)\cdots(\exists f_r)$$
$$\bigl[\,T'\ \wedge\ P'\ \wedge\ f_i\ \text{are bijective mappings}\ \ldots\,\bigr]\ \wedge\ \neg R(y_1,\ldots,y_r,t)$$

leads to a contradiction in $MT_\Sigma$, so that the negation of this relation is a theorem. Because of (4.2.1) $R(y_1, \ldots, y_r, t)$ is a theorem in $MT_\Sigma$, i.e., it can not bring any enrichment compared to the species of structures $\Sigma$. Thus, we have shown that $V$ is not a structure of species richer than $\Sigma$. The relation (4.2.1) can often be proved by taking terms (not necessarily intrinsic terms) $x_1, \ldots, x_n, s$ and $f_1, \ldots, f_n$ in $MT_\Sigma$, such that the relation between brackets of (4.2.1) is a theorem in $MT_\Sigma$.

Thus, for two equivalent species of structures $\Sigma$ and $\Sigma'$, one sees particularly well that $V(x_1, \ldots, x_n, s)$ is a representation of $\Sigma$ in $\Sigma'$ and that $W(x_1, \ldots, x_n, t)$ is a representation of $\Sigma'$ in $\Sigma$. It is enough to pose $E_i(x_1, \ldots) = x_i$ and to choose in (4.2.1) $y_i = x_i$, $f_i$ as identical mappings, and $W(x_1, \ldots, x_n, t) = s$, since $V(x_1, \ldots, x_n, W(x_1, \ldots, x_n, t)) = t$. It follows that (4.2.1) is also a theorem for $W(\cdots)$ and that $W(\cdots)$ is a representation of $\Sigma'$ in $\Sigma$.

A familiar example for a representation of a species of structures $\Sigma$ is analytical geometry. As $MT_{\Sigma'}$ we choose $MT$ (i.e., we introduce no $\Sigma'$ at all). As a single $M_i$ we take $M = \mathbb{R} \times \mathbb{R}$ and as the structure $V \in \mathcal{P}(M \times M \times \mathbb{R})$ the set of all $(\alpha_1, \alpha_2; \beta_1, \beta_2, \delta)$ with

$$\delta = g(\alpha_1, \alpha_2; \beta_1, \beta_2) = \sqrt{(\alpha_1 - \beta_1)^2 + (\alpha_2 - \beta_2)^2}.$$

We determine $\Sigma$ by a principal base set $y$, a structure term $t$ with the typification

$$T(y, t): t \in \mathcal{P}(y \times y \times \mathbb{R}), \quad \text{i.e., } t \subset y \times y \times \mathbb{R},$$

and the axiomatic relation

$P(y, t)$: saying that $t$ determines a function $d : y \times y \to \mathbb{R}_+$, and giving conditions for $d$ that make $y$ a two-dimensional Euclidean space (which is not given in detail here). (See Sects. 4.3 and 4.4.)

Then $V$ is a structure of species $\Sigma$ on the principal base set $M = \mathbb{R} \times \mathbb{R}$, and even a representation of $\Sigma$. In this case, the theorem (4.2.1) becomes

$$(\exists f)\big[\, f : y \to M \text{ is a bijective mapping}$$
$$\text{with } d(z_1, z_2) = g(f(z_1), f(z_2)) \,\big]. \tag{4.2.2}$$

The proof of this theorem is nothing other than the proof that Euclidean geometry $\Sigma$ can be represented by orthogonal coordinates (namely $f : y \to \mathbb{R} \times \mathbb{R}$).

## 4.3 Axiomatic Basis and Fairy Tales

We have seen in Sect. 3.3 that the essential physical structures are hidden in the choice of the picture terms $Q_i$ and $s_\nu$, as terms defined in $MT$. In the usual form of physical theories, one considers only $MT$ and the picture terms without the mathematization $MT_\Theta$ of the connection between mathematics and reality, and one replaces the mathematization $MT_\Theta$ by more or less intuitive methods, similar to the description of "Abbildungsprinzipien" and "Unscharfe Abbildungsprinzipien" in [1]. The theory $MT_{\Sigma(Q_i,s_\nu)}$ is in any case the central part, where the so-called physical laws (or better, the idealized physical laws) are described or hidden. The task will be to change the form of $MT_{\Sigma(Q_i,s_\nu)}$ in such a way that the idealized physical laws appear as axioms.

What are the reasons for which physical laws are usually hidden in $MT_{\Sigma(Q_i,s_\nu)}$? These reasons were already mentioned at the beginning of Chap. 4: A not very deep reason is that in the mathematical formulation $MT_{\Sigma(Q_i,s_\nu)}$, the solution of problems is, in most cases, much simpler than in the form where the physical laws appear as axioms. A much deeper reason is that in $MT$, fairy tales can be told about the reality; one can formulate in $MT$ the description of imagined realities.

This description consists, on the one hand, in a formulation of a mathematical theory (i.e. an $MT$) and, on the other hand, in an interpretation by imagined realities. By such an interpretation one means to show that in $MT$ some of the deductible picture terms $Q_i, s_\nu$ have this real significance, which they have in $MT_\Delta$ by (cor) and the mappings $\phi$.

If we take the example of the Boltzmann theory, we have as $MT$ a mathematical description of Newton's mechanics of $N$ mass points with given forces between these mass points. The mathematical theory $MT$ can be formulated without any imagined interpretation in such a way that a mathematician can investigate this $MT$ with or without any imagined interpretation. If one now takes the imagined interpretation that the mathematical "mass points" are "atoms," i.e., real mass points, then one can intuitively go to the following terms $Q_i, s_\nu$ describing the motion of gas

$$\varrho(\vec{r},t) = m \int f(\vec{r},\vec{v},t)\, \mathrm{d}^3\vec{v}$$

as the mass density of gas,
and

$$\vec{u}(\vec{r},t) = \frac{m}{\varrho(\vec{r},t)} \int \vec{v} f(\vec{r},\vec{v},t)\, \mathrm{d}^3\vec{v}$$

as the velocity of the gas.

Since we will introduce *no* "a priori" *restrictions* to the imagined realities, we will give no other restriction to $MT$ and the terms $Q_i, s_\nu$ than that $MT$

is of the form $MT_\Sigma$, and that the terms $Q_i, s_\nu$ (and the inaccuracy sets $U$) are intrinsic terms relative to $\Sigma$. The authors of this book will not give any prescriptions for the imagined interpretations of the terms introduced by $\Sigma$. We want to leave open $MT_\Sigma$ to *any* figment of the imagination. We also do not want to give any "imagined foundation" of the picture terms $Q_i, s_\nu$ by fairy tales. We only presume that the terms $Q_i, s_\nu$ are defined in $MT_\Sigma$ as intrinsic terms. In any case, the physical theory invented by imagination can, at first, only be tested in the form $(MT_\Sigma)_\Delta$ as described by $(MT_\Sigma)_\Delta \overline{A}$ in Sect. 3.3.2. Whether other terms of $(MT_\Sigma)_\Delta$ can also be interpreted as picture terms describing realities will be dealt with later in Chap. 6.

Since we only know that the terms $Q_i, s_\nu$ are related to the interpreted (by (cor)) terms $\overline{M}_i, \overline{s}_\nu$, and since we do not want to make any assumption that the fairy tales are real, we want to seek for $MT_\Sigma$ another form $MT_{\widehat{\Sigma}}$ where the picture terms $Q_i, s_\nu$ of $MT_\Sigma$ are replaced by the base terms $\widehat{M}_i$ and the structure terms $\widehat{s}_\nu$ of $\widehat{\Sigma}$. We want to find a $\widehat{\Sigma}$ such that the theory $MT_\Delta$ (formed with $MT_\Theta$ and $MT_\Sigma$) is "the same" as the theory $MT_{\widehat{\Delta}}$ (formed with $MT_\Theta$ and $MT_{\widehat{\Sigma}}$).

We reach this aim by choosing a $\widehat{\Sigma}$ such that the terms $Q_i, s_\nu$ form a representation of $\widehat{\Sigma}$ (as defined in Sect. 4.2):

- The species of structures $\widehat{\Sigma}$ is characterized by the principal base sets $\widehat{M}_i$, the structure terms $\widehat{s}_\nu$ with the typification $\widehat{s}_\nu \subset \widehat{S}_\nu(\widehat{M}_1, \ldots)$, and the axiomatic relation $\widehat{P}(\widehat{M}_1, \ldots, \widehat{s}_1, \ldots)$;
- The species of structures $\Sigma$ may be characterized by the principal base sets $X_i$, the structure terms $v_\nu$ with the typification $v_\nu \subset S_\nu(X_1, \ldots)$, and the axiomatic relation $P(X_1, \ldots, v_1, \ldots)$.

The terms $Q_i, s_\nu$ form a representation of $\widehat{\Sigma}$ if the relation (4.2.1), i.e.,

$$(\exists X_1) \cdots (\exists v_1) \cdots (\exists f_1) \cdots$$
$$[\, v_\nu \subset S_\nu(X_1, \ldots)$$
$$\wedge\, P(X_1, \ldots, v_1, \ldots)$$
$$\wedge\, f_i : \widehat{M}_i \to Q_i(X_1, \ldots, v_1, \ldots) \quad \text{are bijective mappings}$$
$$\text{with } \langle f_1, \ldots \rangle^{\widehat{P s_\nu}} \widehat{s}_\nu = s_\nu(X_1, \ldots, v_1, \ldots) \,] \qquad (4.3.1)$$

is a theorem in $MT_{\widehat{\Sigma}}$.

To form $MT_\Delta$ and $MT_{\widehat{\Delta}}$ we need inaccuracy sets. We can expect that $MT_\Delta$ and $MT_{\widehat{\Delta}}$ are the "same" only if we take for $\Delta$ and $\widehat{\Delta}$ the "same" inaccuracy sets. We define the "same" inaccuracy sets by the following correspondence: For bijective mappings $f_i : \widehat{M}_i \to Q_i$ we say that two inaccuracy sets $\widehat{U}$ and $U$ are corresponding sets (relative to the $f_i$) if $U$ is the bijective picture of $\widehat{U}$ by the $f_i$. We can therefore add inside the brackets of (4.3.1) the condition

## 4.3 Axiomatic Basis and Fairy Tales

$f_i : \widehat{U} \to U$ are bijective mappings

for corresponding $\widehat{U}, U$.

We can now form $MT_{\widehat{\Delta}}$ in the same way as $MT_\Delta$ (replacing the terms $Q_i, s_\nu$ by the terms $\widehat{M}_i, \widehat{s}_\nu$). The axiom (3.3.5) takes the form

$$P_{\widehat{\Delta}} : (\exists \widehat{\phi}_1)(\exists \widehat{\phi}_2) \cdots$$
$$[\,\widehat{\phi}_i : \overline{M}_i \to \widehat{M}_i \text{ are injective mappings}$$
$$\text{with } (\widehat{\phi}_i \overline{M}_i)_{\widehat{U}_i} = \widehat{M}_i$$
$$\wedge \widehat{\phi} \overline{s} \subset (\widehat{s})_{\widehat{U}_s}$$
$$\wedge \widehat{\phi} \overline{s}' \subset (\widehat{s}')_{\widehat{U}_s}\,]. \tag{4.3.2}$$

The $\widehat{M}_i$ and $\widehat{s}_\nu$ are auxiliary sets for $\widehat{\Delta}$!

We see immediately that if the terms $\overline{s}_\nu$ over the base $\overline{M}_i$ form a structure of species $\widehat{\Delta}$, then it follows that the terms $\overline{s}_\nu$ over the base $\overline{M}_i$ form a structure of species $\Delta$. One has only to choose for the mappings $\phi_i : \overline{M}_i \to Q_i$, the $\phi_i = f_i \widehat{\phi}_i$ for which the $f_i$ are bijective mappings $\widehat{M}_i \to Q_i$ existing according to the theorem (4.3.1) with (4.3.2). In the same way it follows, conversely, that a structure of species $\Delta$ is also a structure of the $\widehat{\Delta}$. Then we say that $\widehat{\Delta}$ and $\Delta$ are the same species of structures. Thus the tests of $MT_{\widehat{\Delta}}$ or $MT_\Delta$ by $\overline{A}$ are the same.

We call $MT_{\widehat{\Delta}}$ an axiomatic basis of the $PT$ and $\widehat{P}$ the *idealized laws of nature* of this $PT$.

But we have not yet proved that for every $MT_{\Sigma(Q_i, s_\nu)}$ there is an axiomatic basis $MT_{\widehat{\Sigma}(\widehat{M}_i, \widehat{s}_\nu)}$. This is the case if the terms $s_\nu$ over the base $Q_i$ form a structure of species $\widehat{\Sigma}$ and if (4.3.1) is a theorem in $MT_{\widehat{\Sigma}}$. But the terms $s_\nu$ over the base $Q_i$ form a structure of species $\widehat{\Sigma}$ if the axiomatic relation $\widehat{P}(\widehat{M}_1, \ldots, \widehat{s}_1, \ldots)$ transported to the $Q_i, s_\nu$ is a theorem in $MT_\Sigma$ (see Sect. 4.1).

The axiomatic relation $\widehat{P}$ transported to the $Q_i, s_\nu$ is a theorem in $MT_\Sigma$, and (4.3.1) is a theorem in $MT_{\widehat{\Sigma}}$ if we choose, e.g., the relation (4.3.2) as axiomatic relation $\widehat{P}$. In the

### Example A

we get an axiomatic basis $MT_{\widehat{\Sigma}}$ characterized by

1. the principal base set $\widehat{M}$,
2. the structure term $\widehat{s}$,
3. with the typification $\widehat{s} \subset \widehat{M} \times \widehat{M} \times \mathbb{R}$,

4. the axiomatic relation $\widehat{P}$, according to (4.3.1),

$(\exists f)\ [\ f : \widehat{M} \to Q = \mathbb{R} \times \mathbb{R}\ $ is a bijective mapping

with $f\widehat{s} = s\ ]$,

where $s \subset Q \times Q \times \mathbb{R}$ is the set of all $(q_1, q_2, d)$

with $d(\alpha_1, \alpha_2; \beta_1, \beta_2) = \sqrt{(\alpha_1 - \beta_1)^2 + (\alpha_2 - \beta_2)^2}$ (4.3.3)

(see Sect. 3.3.1).

One has to insert in the square brackets that for an $\widehat{U}$ corresponding to $U$ of (3.3.6),

$$f\widehat{U} = U.$$

A fixing of the picture of one $\bar{a}_0$ (as described in Sect. 3.3.2) can also be transported to $MT_{\widehat{\Delta}}$. With $\phi(\bar{a}_0) = (0,0,0)$ we have only to set with $\widehat{\phi}_i \bar{q}_0 = \widehat{a}_0 : f_i(\widehat{a}_0) = (0,0,0)$.

This example reinforces the feeling that the relation (4.3.1) does not give more insight into the structure of reality, since (4.3.1) only says that the structure of the application domain $A_p$ is such that one can imagine realities described by $\Sigma$. But there can be many parts of $\Sigma$ which have no influence on $\widehat{\Sigma}$. For instance, one can add in $\Sigma$ new structures which have no influence on the $Q_i, s_\nu$. Therefore, there remains the question of knowing if it is possible to formulate such an axiomatic relation $\widehat{P}$ that tells us something *only* about the structure of the application domain $A_p$, and not about added fairy tales.

## 4.4 Pure Laws of Nature

The reason for not using fairy tales in the formulation of an axiomatic basis can be explained very well by the

*Example A*

of a two-dimensional Euclidean geometry. The mathematical theory $MT_{\Sigma(Q,s)}$ is given in this example (see end of Sect. 4.3) by $Q = \mathbb{R} \times \mathbb{R}$, $s \subset Q \times Q \times \mathbb{R}$, and the axiomatic relation that $s$ determines a mapping $d : Q \times Q \to \mathbb{R}$ with

$$d(\alpha_1, \alpha_2; \beta_1, \beta_2) = \sqrt{(\alpha_1 - \beta_1)^2 + (\alpha_2 - \beta_2)^2}. \tag{4.4.1}$$

An axiomatic basis is given by $\widehat{\Sigma}(\widehat{M}, \widehat{s})$ with the typification $\widehat{s} \subset \widehat{M} \times \widehat{M} \times \mathbb{R}$ and the axiomatic relation (4.3.3).

If we take (4.3.3) as axiomatic relation $\widehat{P}$, we know that $MT_{\widehat{\Sigma}(\widehat{M},\widehat{s})}$ is a two-dimensional Euclidean geometry. But we want to formulate $\widehat{P}$ in such a way that it immediately says something about $\widehat{s}$, so that we can prove later (4.3.3) as a theorem. Such a system of axioms $\widehat{P}$ can, e.g., begin with the axioms

1. $\widehat{s}$ determines a mapping $\widehat{M} \times \widehat{M} \to \mathbb{R}_+$, i.e., a real function $d(x_1, x_2) \geq 0$,
2. $d(x_1, x_2) + d(x_2, x_3) \geq d(x_1, x_3)$,

and can be extended by additional axioms of a similar form.

In this way we can find a $\widehat{P}$ which "depends" only on $\widehat{M}$ and $\widehat{s}$, i.e., an axiomatic relation where the existence of terms is postulated only if these terms can be "formulated" by $\widehat{M}$ and $\widehat{s}$ and not by the help of terms formulated by real numbers (as, e.g., $\mathbb{R} \times \mathbb{R}$). It is not necessary for the formulation of $\widehat{P}$ to "imagine" terms other than $\widehat{M}$ and $\widehat{s}$.

We will now try to formulate, in general, a condition for the derived form of $\widehat{P}$. Let us denote an axiomatic relation $\widehat{P}$ as *physically interpretable* ("in an idealized form") if $\widehat{P}$ contains an existential quantifier $(\exists z)$ only if $z$ is an element of the echelon set $S(\widehat{M}_1, \ldots, \widehat{s}_1, \ldots)$ of *only* the sets $\widehat{M}_i$ and $\widehat{s}_\nu$ and no other terms. Thus real numbers can only enter in $S(\widehat{M}_1, \ldots, \widehat{s}_1, \ldots)$ indirectly by the $\widehat{s}_\nu$: This means that we can have in $\widehat{P}$ only relations of the form

$$(\exists z)\big[z \in S(\widehat{M}_1, \ldots, \widehat{s}_1, \ldots) \wedge \cdots \big]. \tag{4.4.2}$$

(It is clear that in (4.4.2) $z \in S(\cdots)$ can be omitted if after "$\wedge$" (and) are added relations of which $z \in S(\cdots)$ can be deduced.)

For an axiomatic basis $MT_{\widehat{\Sigma}}$ with a physically interpretable axiomatic relation $\widehat{P}$, we call $\widehat{P}$ also the *idealized pure laws of nature*. Such an axiomatic basis will be briefly called a *simple axiomatic basis*.

The advantage of a simple axiomatic basis is that in $\widehat{\Sigma}$, there appears only terms which are connected by the mapping $\phi_i$ to the terms $\overline{M}_i, \overline{s}_\nu$ describing by (cor) the realities of the application domain $A_p$. "Imagined" terms do not appear in $\widehat{\Sigma}$. But this does not mean that we cannot "add" imagined terms (i.e., fairy tales) by *theorems* (and not axioms!) of the form (4.3.1), and there can be many theorems of this form. For instance, quantum mechanics was developed in two different ways: firstly, by Heisenberg, Born, and Jordan, and secondly, by Schrödinger. Even today discussions about what is the "right" interpretation of quantum mechanics has not been finalized. It cannot be finalized because it is impossible to decide by belief (or by faith!) which fairy tale is true. Fairy tales are, e.g., the so-called "hidden variable" theories and the theory starting from "states" of the systems and a "superposition principle" where the states are represented by vectors in a Hilbert space, and the superposition principle by the mathematical relation of the addition of two vectors. Therefore, a simple axiomatic basis would be very useful since such a basis would describe (even in an idealized form) that part of reality about which we can be sure. Such an axiomatic basis for quantum mechanics is, e.g., described in [8]. "Hidden variable theories" have to prove that they are theorems of the form (4.3.1) in this axiomatic basis. These theories do not claim to prove the reality of such variables; they try to show under which circumstances "imagined" variables are possible. The theories of "states" claim that every micro-system "has" a real state described by a vector in a Hilbert space. Also, if the authors of this book presume that a theorem of the form (4.3.1) was proved (which we have not seen), they cannot believe in the reality of these so-called states, especially since such events as, e.g., the so-called "collapse of wave packets" seems to be unrealistic – a fairy tale.

The problem of the reality of fairy tales of the form (4.3.1) will be considered in Chap. 6. In the context of theorems of the form (4.3.1) one finds in literature the notion of "theoretical concepts." Since we do not believe in an a priori reality of terms appearing in (4.3.1), we will not introduce the designation of these terms as theoretical concepts. We will call these terms only "terms," and nothing else.

The fact that one often designates such terms with words which seem to describe realities does not mean anything. For instance, the word "state" for the vectors in a Hilbert space (in the above-mentioned "state" theory of quantum mechanics) does not say anything more than the word "vector" since the word "state" is not defined and can only be "defined" by the well-defined word "vector" in a Hilbert space. It is a little disappointing if one tries to give the feeling of reality by this word "state," a reality in which one must not believe!

## 4.5 Change of the Mathematical Form of an Axiomatic Basis

The mathematical form of an axiomatic basis can often be changed for the purpose of simplifying the mathematical development of the theories. No deeper physical comprehension is aimed at.

A change of the mathematical form of the theory $MT_{\widehat{\Sigma}}$ which is often used is the change of the axiomatic relation $\widehat{P}$. Let $\widehat{\Sigma}_1$ be the species of structures related to $\widehat{P}_1$, and let $\widehat{\Sigma}_2$ be the species of structures related to $\widehat{P}_2$. If $\widehat{P}_2$ is a theorem in $\widehat{\Sigma}_1$ and $\widehat{P}_1$ is a theorem in $\widehat{\Sigma}_2$, then the theories $MT_{\widehat{\Sigma}_1}$ and $MT_{\widehat{\Sigma}_2}$ are the same theories. The choice of a "simple" axiomatic basis is only one of the various possibilities, i.e., a possibility where $\widehat{P}$ describes directly the (idealized) physical structure of the application domain $A_p$. But it does not mean that this form is such that the *practical* mathematical work is simpler! In our first example of a two-dimensional Euclidean geometry we would certainly prefer to work with the analytic form, i.e., with the picture term $\mathbb{R} \times \mathbb{R}$. In the case of quantum mechanics we have for a simple axiomatic

## 4.5 Change of the Mathematical Form of an Axiomatic Basis

basis a set $X$ of preparation procedures, a set $Y$ of registration procedures, and a relation $s \subset X \times Y \times \mathbb{R}$ describing the probability of a registration $y \in Y$ in the case of a preparation $x \in X$ (see [8]). Nevertheless, we will work with the help of Hilbert's space and the operators in Hilbert's space (see [8]). A simple axiomatic basis has in many cases a physical and not a mathematical advantage; a physical advantage if we are interested in the significance of the axiomatic relation $\widehat{P}$ as an assertion over structures on the reality (see Chap. 6).

But we will not speak in this section of "all" the changes of the mathematical form of $MT_{\widehat{\Sigma}}$ (without limitation to fantasy). We will only speak of a particular change of the form of an axiomatic basis by which we lose nothing of the physical interpretation.

We start with a simple example: Let $\widehat{M}_1, \widehat{M}_2$ be principal base sets of $\widehat{\Sigma}$ (which are also the picture terms of $\overline{M}_1, \overline{M}_2$). Let $\widehat{s} \subset \widehat{M}_1 \times \widehat{M}_2$ be a relation in $\widehat{\Sigma}$. The structure term $\widehat{s}$ determines a mapping $g : \widehat{M}_2 \to \mathcal{P}(\widehat{M}_1)$ in the following manner: For every $y \in \widehat{M}_2$ is defined the set $g(y)$ of all $x$ with $(x, y) \in \widehat{s}$. By $z = g(\widehat{M}_2)$ we get a new structure term $z \subset \mathcal{P}(\widehat{M}_1)$. Then the relation $(x, y) \in \widehat{s}$ is equivalent to the relation $x \in g(y)$.

We consider the case where $g$ is injective. It suffices that there is no contradiction between the relation "$g$ injective" and $\widehat{P}$, because we can then add "$g$ injective" as an axiom, since this cannot be in contradiction to any $\overline{A}$.

If $g$ is injective, the mapping $g : \widehat{M}_2 \to g(\widehat{M}_2) = z$ is a bijective mapping. In this case, we can change the theory $MT_{\widehat{\Sigma}}$ to $MT_{\widehat{\Sigma}_1}$, where $\widehat{\Sigma}_1$ follows from $\widehat{\Sigma}$ by replacing $\widehat{M}_2$ by $z$ and $\widehat{s}$ by the relation $x \in z = g(\widehat{M}_2)$. The connection of $MT_\Theta$ with $MT_{\widehat{\Sigma}_1}$ is given by the mappings $\psi_1 : \overline{M}_1 \to \widehat{M}_1$, $\psi_2 : \overline{M}_2 \to z$, and the condition that an element $(\overline{x}, \overline{y}) \in \overline{s}$ is mapped onto $\psi_1(\overline{x}) \in (\psi_2(\overline{y}))_U$. Here $(\psi_2(\overline{y}))_U \in \mathcal{P}(\widehat{M}_1)$ is $\psi_2(\overline{y}) \in z$ enlarged with the help of an inaccuracy set $U$ for $\widehat{M}_1$. In the new form $MT_{\widehat{\Sigma}_1}$ we lose $\widehat{M}_2$ as base set and replace $\widehat{s}$ by $x \in z$ where $z$ is a new structure term replacing the structure term $\widehat{s}$ of $MT_{\widehat{\Sigma}}$.

The structure term $z \subset \mathcal{P}(\widehat{M}_1)$ is a set of partial sets $s$ of $\widehat{M}_1$. Therefore one often calls the elements of $z$ "properties" of the elements of $\widehat{M}_1$. Thus we see that we can replace the "relation" $\widehat{s} \subset \widehat{M}_1 \times \widehat{M}_2$ by a "set of properties" $z$. In $MT_{\widehat{\Sigma}_1}$ the set $z$ is mostly infinite. Therefore the "properties" as elements of $z$ cannot be introduced as relations of weight 1 in the mathematical language, because we cannot introduce many infinite constants in this mathematical language!

The fact that the introduction of $MT_{\widehat{\Sigma}_1}$ does not change $MT_\Delta$ will be proved in a general form. This general form may be given by the following properties of $MT_{\widehat{\Sigma}}$.

In $MT_{\widehat{\Sigma}}$ it may be possible to divide the terms $\widehat{M}_k$ into two groups $\widehat{M}_{\mu_k}$ and $\widehat{M}_{\nu_i}$ such that there can be defined in $MT_\Sigma$ intrinsic terms $g_i$, the injective mappings

80   4 Species of Structures and Axiomatic Basis of a $PT$

$$g_i : \widehat{M}_{\nu_i} \to T_{\nu_i}(\widehat{M}_{\mu_k}, \ldots, \mathbb{R}), \tag{4.5.1}$$

where the $T_{\nu_i}$ are echelon sets only over the base sets of the group $\widehat{M}_{\mu_k}$.

We want to discuss in one step an additional situation (which is not necessarily the case).

In the theory $MT_\Sigma$ with the picture terms $Q_i, s_\nu$ (where $\Sigma$ can describe a fairy tale) which form a representation of $\widehat{\Sigma}$, the $g_i$ can be transported as mappings

$$\tilde{g}_i : Q_{\nu_i} \to T_{\nu_i}(Q_{\mu_k}, \ldots, \mathbb{R}). \tag{4.5.2}$$

We will later presume that

the $\tilde{g}_i$ are the identical mappings $Q_{\nu_i} \to Q_{\nu_i}$. \hfill (4.5.3)

This implies

$$Q_{\nu_i} \subset T_{\nu_i}(Q_{\mu_k}, \ldots, \mathbb{R}). \tag{4.5.4}$$

In general (i.e., without (4.5.3) and (4.5.4)) it follows from (4.5.1) that we can construct a new species of structures $\widehat{\Sigma}_1$ which contains as base terms only the group $\widehat{M}_{\mu_k}$. Instead of the terms $\widehat{M}_{\nu_i}$ we introduce in $\widehat{\Sigma}_1$, in addition to the structure terms $\hat{s}_\nu$ (of $\widehat{\Sigma}$), the terms $g_i(\widehat{M}_{\nu_i})$ as structure terms. To define precisely the species of structures $\widehat{\Sigma}_1$, we introduce new letters: instead of the $\widehat{M}_{\mu_k}$ we write $\widehat{N}_k$, instead of the $\widehat{M}_{\nu_i}$ we write $t_i$, and instead of the $\hat{s}_\nu$ we write $t_\nu^{(0)}$. The base terms of $\widehat{\Sigma}_1$ are the $\widehat{N}_k$, the structure terms are the $t_i$ and $t_\nu^{(0)}$. As typification of the structure terms we take

$$t_\nu^{(0)} \subset \widehat{S}_\nu(\widehat{N}_k, T_{\nu_i}(\widehat{N}_k, \ldots, \mathbb{R})) \tag{4.5.5}$$

with the $\widehat{S}_\nu$ of the axiomatic basis $\widehat{\Sigma}$ (see Sect. 4.3), but by replacing the $\widehat{M}_{\mu_k}$ by the $\widehat{N}_k$ and the $\widehat{M}_{\nu_i}$ by the $T_{\nu_i}(\widehat{N}_k, \ldots, \mathbb{R})$ with $T_{\nu_i}$ from (4.5.1).

For the typification of the $t_i$ we take

$$t_i \subset T_{\nu_i}(\widehat{N}_k, \ldots, \mathbb{R}). \tag{4.5.6}$$

Now we have to introduce the axiomatic relation $\widehat{P}_1(\widehat{N}_k, t_i, t_\nu^{(0)})$ for $\widehat{\Sigma}_1$. We compose $\widehat{P}_1$ with three relations:

- At first we take the axiomatic relation $\widehat{P}_{01}(\widehat{N}_k, t_i, t_\nu^{(0)})$, which we get from $\widehat{P}(\widehat{M}_i, \hat{s}_\nu)$ by replacing the $\widehat{M}_{\mu_k}$ by the $\widehat{N}_k$, the $\widehat{M}_{\nu_i}$ by the $t_i$, and the $\hat{s}_\nu$ by the $t_\nu^{(0)}$.

## 4.5 Change of the Mathematical Form of an Axiomatic Basis    81

- As second relation we add

$$t_\nu^{(0)} \subset \widehat{S}_\nu(\widehat{N}_k, t_i), \tag{4.5.7}$$

which we get from (4.5.5) by replacing the $T_{\nu_i}(\widehat{N}_k, \ldots, \mathbb{R})$ by the $t_i$. Together with (4.5.6) it follows that we can leave out (4.5.5).
- The third relation is obtained as follows: From (4.5.1) and the first axiomatic relation $\widehat{P}_{01}$, it follows that there are intrinsic terms $\widehat{g}_i$ the injective mappings

$$\widehat{g}_i : t_i \to T_{\nu_i}(\widehat{N}_k, \ldots, \mathbb{R}). \tag{4.5.8}$$

As an additional axiom we introduce that the $\widehat{g}_i$ are the identical mappings $t_i \to t_i$.

We see immediately that the terms $t_\nu^{(0)}$ over the base $\widehat{N}_k, t_i$ form a representation of $\widehat{\Sigma}$. From the axiomatic relation $\widehat{P}_1$ and (4.5.1) it even follows that the terms $t_\nu^{(0)}$ give a representation of $\widehat{\Sigma}$ in $MT_{\widehat{\Sigma}_1}$. Before proving this, we want to introduce a species of structures $\widehat{\Delta}_1$ starting from the same $MT_\Theta$ and taking as picture terms instead of the terms $\widehat{M}_i$ of $MT_{\widehat{\Sigma}}$ the terms $\widehat{N}_k, t_i$ of $MT_{\widehat{\Sigma}_1}$, and instead of the structure terms $\widehat{s}_\nu$ the terms $t_\nu^{(0)}$ (see the introduction of $\Delta$ in Sect. 3.3.1). If the terms $t_\nu^{(0)}$ over the base $\widehat{N}_k, t_i$ form a representation of $\widehat{\Sigma}$ in $MT_{\widehat{\Sigma}_1}$, then the two species of structures $\widehat{\Delta}$ and $\widehat{\Delta}_1$ are the "same" (see Sect. 4.3).

Since the terms $\widehat{N}_k, t_i, t_\nu^{(0)}$ of $\widehat{\Sigma}_1$ are interpreted by $\widehat{\Delta}_1$, we also call $MT_{\widehat{\Sigma}_1}$ an axiomatic basis. It is necessary to observe that the picture terms for the $\widehat{M}_i$ are not only the $\widehat{N}_k$ but also the $t_i$! If one does not observe this, one can make mistakes.

An axiomatic basis $MT_{\widehat{\Sigma}}$ is called an *axiomatic basis of the first order*, and an axiomatic basis $MT_{\widehat{\Sigma}_1}$ is called an *axiomatic basis of a higher order* since the typification of the $t_i$ contains the procedure $\mathcal{P}$.

Now we want to prove that the terms $t_\nu^{(0)}$ over the base $\widehat{N}_k, t_i$ form a representation of $\widehat{\Sigma}$ in $MT_{\widehat{\Sigma}_1}$, i.e., we have to prove that the relation (4.2.1) is a theorem in $MT_{\widehat{\Sigma}}$. The relation (4.2.1) has the form

$$(\exists \widehat{N}_1) \cdots (\exists t_1) \cdots (\exists t_1^{(0)}) \cdots (\exists f_1) \cdots$$

$[\; t_i \;\subset T_{\nu_i}(\widehat{N}_k, \ldots, \mathbb{R})$

$\land \; t_\nu^{(0)} \subset \widehat{S}_\nu(\widehat{N}_k, t_i)$

$\land \; \widehat{P}_{01}(\widehat{N}_k, t_i, t_\nu^{(0)})$

$\land \; f_1, \ldots$ are bijective mappings $f_{\mu_k} : \widehat{M}_{\mu_k} \to \widehat{N}_k$, $f_{\nu_i} : \widehat{M}_{\nu_i} \to t_i$

with $\langle f_1, \ldots \rangle^{P\widehat{S}_\nu} \widehat{s}_\nu = t_\nu^{(0)}$ ]. (4.5.9)

To show that (4.5.9) is a theorem in $MT_{\widehat{\Sigma}}$, we only need to set $\widehat{N}_k = \widehat{M}_{\mu_k}$, to set $f_{\mu_k} : \widehat{M}_{\mu_k} \to \widehat{N}_k$ as the identical mappings $\widehat{M}_\mu \to \widehat{M}_\mu$, to set $f_{\nu_i} : \widehat{M}_{\nu_i} \to t_i$ equal to the $g_i$ of (4.5.1), and to define the $t_\nu^{(0)}$ by $\langle f_1, \ldots \rangle^{P\widehat{S}_\nu} \widehat{s}_\nu$.

In a similar way, it is possible to prove that the terms $g_i(\widehat{M}_{\nu_i}), \tilde{t}_\nu$

(where

$$\tilde{t}_\nu = \langle f_1, \ldots \rangle^{P\widehat{S}_\nu} \widehat{s}_\nu \qquad (4.5.10)$$

with $f_1, \ldots$ are identical mappings $f_\mu : \widehat{M}_\mu \to \widehat{M}_\mu$ and $f_{\nu_i} = g_i$)

over the base $\widehat{M}_{\mu_k}$ form a representation of $\widehat{\Sigma}_1$ in $MT_{\widehat{\Sigma}}$.

Now we want to show how these two species of structures $\widehat{\Sigma}$ and $\widehat{\Sigma}_1$ and the relation between these two structures reflect in $MT_\Sigma$.

We start with the fact that the terms $Q_i, s_\nu$ form a representation of $\widehat{\Sigma}$ in $MT_\Sigma$, and we assume, according to (4.5.3), that the mappings $\tilde{g}_i$ are the identical mappings $Q_{\nu_i} \to Q_{\nu_i}$. Here the terms $Q_i$ are divided corresponding to the $\widehat{M}_k$ into two groups $Q_{\mu_k}$ and $Q_{\nu_i}$. We want to prove that the terms $Q_{\nu_i}, s_\nu$ over the base $Q_{\mu_k}$ form a representation of $\widehat{\Sigma}_1$.

To begin with we will prove that the terms $Q_{\nu_i}, s_\nu$ over the base $Q_{\mu_k}$ form a structure of species $\widehat{\Sigma}_1$ if the terms $s_\nu$ over the base $Q_i$ form a structure of species $\widehat{\Sigma}$. This is the case, if the relation

$$\begin{aligned}
&[ \, Q_{\nu_i} \subset T_{\nu_i}(Q_{\mu_k}, \ldots) \\
&\wedge \, s_\nu \subset S_\nu(Q_1, \ldots) \\
&\wedge \, \widehat{P}_1(Q_{\mu_k}, Q_{\nu_i}, s_\nu) \\
&\wedge \, \tilde{g}_i : Q_{\nu_i} \to Q_{\nu_i} \text{ are identical mappings } ]
\end{aligned} \qquad (4.5.11)$$

is a theorem in $MT_\Sigma$. This follows from (4.5.3) and from the fact that the terms $s_\nu$ over the base of all $Q_i$ form a structure of species $\widehat{\Sigma}$.

The inverse of this theorem holds: If the terms $Q_{\nu_i}, s_\nu$ over the base $Q_{\mu_k}$ form a structure of species $\widehat{\Sigma}_1$, then the terms $s_\nu$ over the base $Q_i$ form a structure of species $\widehat{\Sigma}$.

To prove this theorem we have to prove in $MT_\Sigma$ the relation

$$s_\nu \subset \widehat{S}_\nu(Q_1, \ldots) \wedge \widehat{P}(Q_i, s_\nu). \qquad (4.5.12)$$

Since the terms $Q_{\nu_i}, s_\nu$ form a structure of species $\widehat{\Sigma}_1$, the relation (4.5.11) is a theorem in $MT_\Sigma$ that leads to (4.5.12).

## 4.5 Change of the Mathematical Form of an Axiomatic Basis

If the terms $s_\nu$ over the base $Q_i$ form a representation of $\widehat{\Sigma}$ in $MT_\Sigma$, then also the terms $Q_{\nu_i}, s_\nu$ over the base $Q_{\mu_k}$ form a representation of $\widehat{\Sigma}_1$ in $MT_\Sigma$. To prove this, one has to prove that the relation

$$(\exists X_1)\cdots(\exists v_1)\cdots(\exists h_1)\cdots$$
$$[\,v_\nu \subset S_\nu(X_1,\ldots)$$
$$\wedge\ P(X_1,\ldots,v_1,\ldots)$$
$$\wedge\ h_k : \widehat{N}_k \to Q_{\mu_k}(X_1,\ldots) \quad \text{are bijective mappings}$$
$$\text{with} \quad \langle h_1,\ldots\rangle^{PT_{\nu_i}} t_i = Q_{\nu_i}(X_1,\ldots)$$
$$\wedge\ \langle h_1,\ldots\rangle^{P\widehat{S}_\nu} t_\nu^{(0)} = s_\nu\,] \tag{4.5.13}$$

is a theorem in $MT_{\widehat{\Sigma}}$ (the letters for $MT_\Sigma$ are selected as in (4.3.1)).

Since the terms $s_\nu$ over the base $Q_i$ form a representation of $\widehat{\Sigma}$, we have the theorem (4.3.1). Since the terms $t_\nu^{(0)}$ over the base $\widehat{N}_k, t_i$ form a structure of species $\widehat{\Sigma}$ in $MT_{\widehat{\Sigma}_1}$, we can transport (4.3.1) to $MT_{\widehat{\Sigma}_1}$:

$$(\exists X_1)\cdots(\exists v_1)\cdots(\exists f_1)\cdots$$
$$[v_\nu \subset S_\nu(X_1,\ldots)$$
$$\wedge\ P(X_1,\ldots,v_1,\ldots)$$
$$\wedge\ f_1,\ldots \text{ are bijective mappings } f_{\mu_k} : \widehat{N}_k \to Q_{\mu_k},\ f_{\nu_i} : t_i \to Q_{\nu_i}$$
$$\text{with } \langle f_1,\ldots\rangle^{P\widehat{S}_\nu} t_\nu^{(0)} = s_\nu\,]. \tag{4.5.14}$$

If we replace here the $f_{\mu_k}$ by the $h_k$ and the $f_{\nu_i}$ by the $k_i$, then we get the relation (4.5.13) from (4.5.14) if we prove $k_i t_i = \langle h_1,\ldots\rangle^{PT_{\nu_i}} t_i$.

Since the $g_i$ in (4.5.1) are intrinsic terms, for bijective mappings $f_{\nu_i}$ with $\widetilde{g}_i$ in (4.5.2) we get the following commutative diagram

$$\begin{array}{ccc} Q_{\nu_i} & \xrightarrow{\widetilde{g}_i} & T_{\nu_i}(Q_{\mu_k},\ldots) \\ \uparrow f_{\nu_i} & & \uparrow \langle f_{\nu_i},\ldots\rangle^{T_{\nu_i}} \\ \widehat{M}_{\nu_i} & \xrightarrow{g_i} & T_{\nu_i}(\widehat{M}_{\mu_k},\ldots) \end{array} \tag{4.5.15}$$

From (4.5.15) it follows

$$\widetilde{g}_i f_{\nu_i} = \langle f_{\nu_i},\ldots\rangle^{T_{\nu_i}} g_i.$$

With (4.5.3) it follows

$$f_{\nu_i} = \langle f_{\nu_i},\ldots\rangle^{T_{\nu_i}} g_i,$$

which can be transported to $MT_{\widehat{\Sigma}_1}$ in the form

$$k_i = \langle h_1, \ldots \rangle^{T_{\nu_i}} \tilde{g}_i.$$

Since $\tilde{g}_i$ is on $t_i$ the identical mapping, we get

$$k_i t_i = \langle h_1, \ldots \rangle^{PT_{\nu_i}} t_i. \tag{4.5.16}$$

Thus (4.5.13) is proved.

The inverse is also valid: If the terms $Q_{\nu_i}, s_\nu$ over the base $Q_{\mu_k}$ form a representation of $\widehat{\Sigma}_1$, then the terms $s_\nu$ over the base $Q_i$ form a representation of $\widehat{\Sigma}$.

To prove this, we have to demonstrate that from the relation (4.5.13) as a theorem in $MT_{\widehat{\Sigma}_1}$ it follows the relation (4.3.1) as a theorem in $MT_{\widehat{\Sigma}}$. Since the terms $g_i(\widehat{M}_{\nu_i}), \tilde{t}_\nu$ over the base $\widehat{M}_{\mu_k}$ form a structure of species $\widehat{\Sigma}_1$ in $MT_{\widehat{\Sigma}}$, from (4.5.13) it follows that

$(\exists X_1) \cdots (\exists v_1) \cdots (\exists h_1) \cdots$
$[\ v_\nu \subset S_\nu(X_1, \ldots)$
$\wedge\ P(X_1, \ldots, v_1, \ldots)$
$\wedge\ h_k : \widehat{M}_k \to Q_{\mu_k}(X_1, \ldots, v_1, \ldots)$ are bijective mappings
with $\quad \langle h_1, \ldots \rangle^{PT_{\nu_i}} g_i(\widehat{M}_{\nu_i}) = Q_{\nu_i}(X_1, \ldots, v_1, \ldots)$
$\wedge\ \langle h_1, \ldots \rangle^{P\widehat{S}_\nu} \tilde{t}_\nu = s_\nu(X_1, \ldots, v_1, \ldots)\ ].$

With $f_{\nu_1} = \langle h_1, \ldots \rangle^{PT_{\nu_i}} g_i$ and with the definition $\tilde{t}_\nu = \langle f_1, \ldots \rangle^{P\widehat{S}_\nu} \widehat{s}_\nu$ we get (4.3.1).

We have demonstrated the different mathematical forms $MT$ of the "same" physical theory $PT$:

$$PT \equiv A_p(\text{cor})MT_\Delta \quad \cdots \quad MT.$$

For all various forms of $MT$ we have the "same" $\Delta$ and thus the "same" physics. The various forms can make the "work with the $PT$" more and more fruitful, i.e., make of the $PT$ a tool increasingly more efficient.

Also, a fairy tale $MT_\Sigma$ is without any restrictions. The only condition is that one defines in $MT_\Sigma$ the picture terms $Q_{\nu_i}, s_\nu$. One gets the physically essential mathematical theory $MT_\Sigma$ only by these picture terms.

If someone takes a fairy tale $MT_\Sigma$ without the definition of picture terms $Q_{\nu_i}, s_\nu$, he violates the methods of a physical theory established here. This violation cannot be eliminated by using physically sounding words for some of the terms of $\Sigma$. We call such an $MT_\Sigma$ without the definition of the picture terms an "uninterpreted theory" (see also Sect. 4.9).

The next section will not establish additional conditions for the "method of a physical theory" but will describe some general structures of this method.

## 4.6 Inaccuracy Sets and Uniform Structures

In this section we will provide a very useful mathematical method to describe a set of "possible" inaccuracy sets, i.e., a set out of which we can seek the best (the smallest) inaccuracy sets for which the theory $MT_\Delta$ is not in contradiction with experiment.

Thus the introduction of such a set of possible inaccuracy sets does not describe a new physical reality, but is only a method for obtaining the best description of reality by $MT_\Delta$ when "we fix" the idealized description by the picture terms of $MT$.

Let $X$ be a set in $MT$ such that $X = Q$ for a picture set $Q$, or $s \subset X$ for a picture relation $s$. In such a set $X$ we have to introduce an inaccuracy set $U \subset X \times X$ as described in Sect. 3.3.1. The structure $\Delta$ depends on this inaccuracy set $U$. If we want to stress this dependency, we will write explicitly $\Delta_U$.

For such a set of possible inaccuracy sets we take a set $N \subset \mathcal{P}(X \times X)$ for which the following relations are axioms or theorems of $MT_\Sigma$:

$$\emptyset \neq N \subset \mathcal{P}(X \times X), \tag{4.6.1}$$

$$U_1 \in N \wedge U_2 \subset \mathcal{P}(X \times X) \wedge U_1 \subset U_2 \Rightarrow U_2 \in N, \tag{4.6.2}$$

$$U_1 \in N \wedge U_2 \in N \Rightarrow U_1 \cap U_2 \in N, \tag{4.6.3}$$

$$U \in N \Rightarrow D(X) \subset U, \tag{4.6.4}$$

$$U \in N \Rightarrow U^{-1} \in N \quad (U^{-1} = \{(x,y)|(y,x) \in U\}), \tag{4.6.5}$$

$$U \in N \Rightarrow (\exists U_1)(U_1 \in N \wedge U_1 \circ U_1 \subset U). \tag{4.6.6}$$

The set $N$ with these relations as axioms is a species of structures of $X$; it is called *species of uniform structures*. We take such a structure as a set of *possible inaccuracy sets*. "Possible" means that we can construct with such a set $U \in N$ the theory $MT_{\Delta_U}$, and can attempt to see if this theory is not in contradiction with experiment (i.e., that $MT_{\Delta_U} \overline{A}$ is without contradiction).

Why do we take the above axioms as properties of such a set $N$ of possible inaccuracy sets?

We see immediately that $\Delta_{U_1}$ is richer than $\Delta_{U_2}$ if $U_1 \subset U_2$. Thus $MT_{\Delta_{U_2}} \overline{A}$ is without contradiction if $MT_{\Delta_{U_1}} \overline{A}$ is without contradiction. Therefore we take (4.6.2) as a property of $N$.

If the inaccuracy sets $U_1, U_2$ do not lead to contradictions with experiment, it does *not* follow that $U_1 \cap U_2$ does not lead to contradiction with experiment. Similarly, it does not follow that an inaccuracy set $U_1$ with $U_1 \circ U_1 \subset U$ does not lead to contradiction with experiment if $U$ does not lead to contradiction with experiment. Therefore we take (4.6.3) and (4.6.6) as properties of $N$ for the purpose of formulating methods to obtain *smaller* and smaller inaccuracy sets in order to test these smaller inaccuracy sets, and to see whether they lead to contradiction with experiment.

The property (4.6.4) is obvious since we cannot distinguish an $x \in X$ from itself. The property (4.6.5) is added since we do not want to distinguish by an inaccuracy set $U$ the pair $(x, y)$ from the pair $(y, x)$ (see the introduction of an inaccuracy set in Sect. 3.3.1).

In this sense, a uniform structure $N$ of $X$ describes a "procedure" to get smaller and smaller inaccuracy sets for the purpose of testing whether $\Delta_U$ does not lead to contradiction with experiment. But the axioms (4.6.1) to (4.6.6) give *no* restriction to the inaccuracy sets $U \in N$, since they are fulfilled by the set

$$N_0 = \{U \mid U \subset X \times X \ \wedge \ (4.6.4)\}. \tag{4.6.7}$$

The axioms (4.6.1) to (4.6.6) therefore describe not a "real" structure but only a "physical" structure; a physical procedure to select finer and finer, but still usable, inaccuracy sets. This procedure can only be done "step by step" and not with infinitely many elements of $N$. The description of such a step-by-step procedure is not yet included in the axioms (4.6.1) to (4.6.6), i.e., we have to introduce additional axioms for $N$.

Before we do this, we will make some remarks about the necessary experiments in order to select a usable inaccuracy set; this is the price we pay for the idealization. This experimental work, of obtaining a survey of the usable subset of $N$, is typical in physics and makes physics for some people an "inexact" science. But this is not the case: the idealization $MT$ is "inexact," but it relates by $MT_\Delta$ (for a particular $U$) to an exact assertion about reality, even if this assertion is not "categorical," i.e., the structure $\Delta$ is in any case, because of $U$, not a categorical structure (an "univalent" category, i.e., an "univalent" species of structures). This does not mean that the reality is not categorical, but it means that we do not *know* exactly the structure of the reality, i.e., that we know only an "interval" of structures which are perhaps realized. It is clear that this situation is a stimulation to seek a "better" theory (see Chap. 6).

These considerations could lead to the opinion that the "best" theory should be a categorical theory, but this is a mistake. The world is not determined by a "part" of the world, e.g., by that part that has taken place until "now." A categorical theory would only represent the reality of the "total" world, but if we knew the "total" reality we would not need any theory whatsoever. A theory will tell us something about the reality that we have

## 4.6 Inaccuracy Sets and Uniform Structures

not observed. Since as human beings we can decide by our free will over a part of reality in our future, a physical theory cannot be categorical since a theory must leave open some "possibilities" about the reality. The question of how we can read from a theory these "possibilities" will be elaborated in Chap. 6. For the moment, we will only emphasize that we have to expect that the idealized part $MT_{\widehat{\Sigma}}$ of most of the physical theories will *not* have a categorical species of structure $\widehat{\Sigma}$.

(There are only some idealized categorical theories of space–time: "Galileo–Newton's theory" is an example of such a categorical theory, but also the "special relativity theory." "Einstein's gravitational theory" is not a categorical theory of space–time since we can, e.g., change this structure a little by our own free will. Newton's mechanics is also not categorical since we have at our disposal the "initial values").

As we have mentioned above, the aim of introducing $N$ is to select from $N$ the set of *usable inaccuracy sets* by experiments. But this selection depends on the "measuring errors" which come from the pre-theories (see Chap. 6). If these "errors" are very large, it can be that all elements of $N$ are allowed as inaccuracy sets; the measuring errors do not allow us to detect where the idealization of $MT_\Sigma$ deviates from the reality. In this case we often say that the idealized theory $MT_\Sigma$ is "good enough."

After these general remarks about the inaccuracy sets and the experimental work, we will continue to restrict our "reservoir" $N$ of inaccuracy sets. The biggest reservoir $N_0$ (4.6.7) contains "all" sets which can serve as inaccuracy sets, but it contains so many sets that it is not possible to seek systematically and with a positive result a $U$ for $\Delta_U$. We have already seen above that we can test only "step by step." Therefore this testing is not "fishing in troubled water" if there is in $N$ a countable subset generating the total $N$, i.e., if there is a countable base of $N$. (For the concept of a countable basis, see [9].) We therefore restrict the uniform structure $N$ of $X$ by the axiom

there is a countable base of $N$, (4.6.8a)

which is equivalent to

$X$ is "metrizable" relative to $N$. (4.6.8b)

(For the concept of metrizable, see [10].) The procedure to look for usable inaccuracy sets only makes sense if we have a "reservoir" $N$ which fulfills the axioms (4.6.1) to (4.6.6) and (4.6.8).

Also if these axioms are fulfilled, $N$ can contain elements that are not usable as inaccuracy sets, since they do not satisfy the property (4.3.4) of inaccuracy sets. We therefore introduce for $N$ the following axiom:

all $U \in N$ satisfy (3.3.4) (4.6.9)

By (4.6.9) we remove from $N$ all elements that have no use in inaccuracy sets. This removing does not describe the procedure of searching, but it describes the "direction" in which we have to cancel the idealization of the set $X$, i.e., it suggests a reflection over the difference between reality and idealization.

If the axiom (4.6.1) is valid, then it follows that the set $X$ endowed with the uniform structure $N$ is "precompact" (for the concept of precompact, see [9]).

The uniform structure $N$ of $X$ allows us to complete $X$ to a set $\widetilde{X}$ whereby is defined a canonical mapping $i : X \to \widetilde{X}$. If $\bigcap_{U \in N} U = D(X)$, then $i$ is injective and one can identify $X$ with $iX$, i.e., one can define $X$ as a subset of $\widetilde{X}$. If $i$ is not injective, then there would be elements in $X$ which can "never" be distinguished by experiments. Therefore we can replace $X$ by $iX$. For a physical theory, we therefore always assume that $i$ is injective.

If $X$ is precompact, then $\widetilde{X}$ is compact. If $X$ is compact and metrizable, then it is also separable. One has with the metric $d(x_1, x_2)$ for every $d(x_1, x_2) \leq \varepsilon_\nu$ a finite set $\varepsilon_\nu$-dense in $\widetilde{X}$ and therefore a countable set dense for every $\varepsilon$.

Thus we get the following postulate for picture sets $X$:

For every such set $X$ is defined a uniform structure $N$ such that $X$ is precompact (or compact) and metrizable. Then $X$ is also separable.

This postulate includes also the case that $X$ is a finite set. One has only to take as $N$ the set of all subsets of $X \times X$ which include the diagonal set of all $(x, x)$.

Also, for the sets $\mathbb{R}$ in $s(\ldots, \mathbb{R})$ which are used for the picture sets of measurements, we have to choose an $N$, e.g., by the metric

$$d(\alpha_1, \alpha_2) = \bigl| \operatorname{arctg}(\alpha_1/\alpha_0) - \operatorname{arctg}(\alpha_2/\alpha_0) \bigr|$$

with a fixed number $\alpha_0$.

*Example A*

The inaccuracy set $U_{\delta\varrho}$ given in Sect. 3.3.1 is an element of this $N$ generated by $d(\alpha_1, \alpha_2)$. With this set $N$ the set $\mathbb{R}$ is precompact.

The completion $\widetilde{\mathbb{R}}$ of $\mathbb{R}$ is $\mathbb{R}$ plus two elements, one in $+\infty$ and the other in $-\infty$. It is clear that we have to choose out of this $N$ different inaccuracy sets $U$ for different $s(\ldots, \mathbb{R})$. Also for the set $Q = \mathbb{R} \times \mathbb{R}$ (see Sect. 3.3.1) we can find a metric (which generates a set $N$ of which (3.3.6) is an element) by the following procedure (see Fig. 4.1).

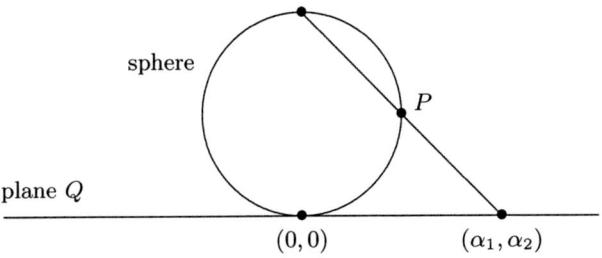

**Fig. 4.1.** Stereographic projection

Every point $(\alpha_1, \alpha_2)$ is mapped onto a point $P$ on the surface of a sphere. As a metric on $Q$ we use the distance of the corresponding points on the surface of the sphere. With this metric, $Q$ is precompact. $\widetilde{Q}$ is the set $Q$ plus one point in the "infinity" of $Q$.

But we can also use another metric with another $\widetilde{Q}$, taking the mapping of Fig. 4.2.

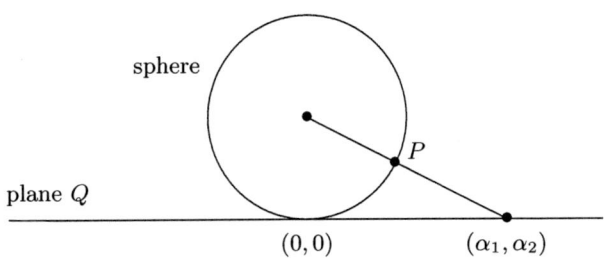

**Fig. 4.2.** Gnomonic projection

The set $N$ contains the set (3.3.6) as an element. For this set $N$, the set $\widetilde{Q}$ is the set $Q$ plus different points in "infinity." But for different directions from $(0,0)$ to "infinity" there also different points.

This example demonstrates that the set $N$ selected is not determined by reality. The set $N$ selected has also something to do with our intention to get such inaccuracy sets $U$ for which $\Delta$ generates a "good" physical theory.

In our example $Q = \mathbb{R} \times \mathbb{R}$ we can also choose a third uniform structure $N$, namely, generated by the distance

$$d(\alpha_1, \alpha_2; \beta_1, \beta_2) = \sqrt{(\alpha_1 - \beta_1)^2 + (\alpha_2 - \beta_2)^2}.$$

With this metric, $Q$ is not precompact! Therefore, this set $N$ is not so efficient in *seeking* usable inaccuracy sets. $U$ from (3.3.6) is an element of $N$, but this $N$ contains too many elements to find anything.

Since in many physical theories we use mathematical idealizations endowed with different uniform structures for the same sets, we will add here some remarks about this phenomenon.

In most physical theories, the introduction of idealized sets $X$ is connected to the introduction of a metric which tells us what is "small" and what is "large." For example, small and large differences of spatial positions, small and large differences of time, small and large volumes, small and large velocities, small and large electric fields, etc. The idealizations consist of the mathematical formulation that this "small" can be made smaller and smaller without any finite limit, and that this "large" can be made larger and larger without any finite limit. But the opinion of the authors is such that the extrapolations are not real. They are only fictive extrapolations made because we do *not know* how the reality is for small and large regions. To take these extrapolations as real is therefore only a swindle, by which we try to change our ignorance into knowledge. Instead of this swindle, we have to choose finite inaccuracy sets and for this purpose to find a suitable uniform structure $N$ with such properties as described above.

A way of finding such an $N$ is, e.g., the following method which is often used:

We choose real functions $X \to \mathbb{R}$ which do not change "too fast" in the small and in the large. We begin with the description of this method step by step.

At first we consider a set of real functions $f_\lambda : X \to \mathbb{R}$ (where $\lambda$ is an element of an index set $s$). We only take functions for which $|f_\lambda(x)| < \varrho$ for all $\lambda$ and $x$, i.e., the $f_\lambda$ maps $X$ onto a bounded region of $\mathbb{R}$, i.e., onto a precompact region of $\mathbb{R}$ (precompact relative to the usual uniform structure of $\mathbb{R}$). The set of functions $f_\lambda$ will describe how well we can distinguish the various $x$ by measurements. Indeed the set of functions $f_\lambda$ determines an initial uniform structure $N$ (see [9]), which describes this distinguishability. Since all $f_\lambda$ map $X$ onto a precompact region, $N$ is precompact (see [9]).

If it is possible to select from the set of the $f_\lambda$ a countable subset $f_{\lambda_k}$ such that the uniform structure generated by the subset $f_{\lambda_k}$ is the same as the uniform structure generated by all $f_\lambda$, then $X$ is metrizable (see [10]). The following condition is sufficient for such a set of $f_{\lambda_k}$:

To every $f_\lambda$ and to every $\varepsilon > 0$ there is a $f_{\lambda_k}$ for which $|f_\lambda(x) - f_{\lambda_k}(x)| < \varepsilon$ for all $x \in X$.

This condition is sufficient if we prove that all $f_{\lambda_k}(x)$ are uniformly continuous for the initial uniform structure generated by the $f_{\lambda_k}$. This follows from

## 4.6 Inaccuracy Sets and Uniform Structures

$$|f_\lambda(x_1) - f_\lambda(x_2)| \leq |f_\lambda(x_1) - f_{\lambda_k}(x_1)|$$
$$+ |f_{\lambda_k}(x_1) - f_{\lambda_k}(x_2)|$$
$$+ |f_{\lambda_k}(x_2) - f_\lambda(x_2)|$$
$$\leq 2\varepsilon + |f_{\lambda_k}(x_1) - f_{\lambda_k}(x_2)|.$$

We will now use this theorem for the consideration of a widespread mathematical structure in physical theories.

Let $p$ and $g$ be two uniform structures on $X$. $X_p$ may be $X$ endowed with the structure $p$, and $X_g$ may be $X$ endowed with the structure $g$. We consider the case where $g$ is finer than $p$, but the topology generated by $g$ and $p$ may be equal. Let $\widetilde{X}_p$ and $\widetilde{X}_g$ be the completions of $X_p$ and $X_g$ respectively. Since $g$ is finer than $p$, the canonical injection $i : X_g \to \widetilde{X}_p$ is uniformly continuous relative to the structure $p$ and can therefore be continued to $i : \widetilde{X}_g \to \widetilde{X}_p$ (see [9]). By $i$ is then, as an initial uniform structure on $\widetilde{X}_g$, determined a uniform structure which coincides on the subset $j(X) \subset \widetilde{X}_g$ (with $j$ as the canonical mapping $X \to \widetilde{X}_g$) with the uniform structure $p$. $\widetilde{X}_p$ can therefore also be considered as the completion of $\widetilde{X}_g$ relative to the uniform structure $p$. In order not to distinguish in such a case between $X, \widetilde{X}_g, \widetilde{X}_p$, we take as picture term $X$ the term $\widetilde{X}_g$, i.e., we can assume in such cases, for simplicity, that $X = \widetilde{X}_g$, i.e., $X$ is complete (and separable) relative to $g$. Since we assume that $\widetilde{X}_p$ is separable and the topology of $p$ and $g$ coincides on $X$, we also get that $X_g$ is separable.

Since on a compact space the uniform structure is determined by the topological structure (see [9]), we get that on a compact subset of $X_g$ the uniform structures $g$ and $p$ are identical, i.e., that every compact subset of $X_g$ is also complete as a subset of $X_p$.

Since in physical theories we only have the case that $X_g$ is metrizable, we want to prove at the end of this section the following theorem.

If $X_g$ is separable and metrizable, then there is a uniform structure $p$, for which $X_y$ is precompact, metrizable, and separated, and such that the topologies of $X_p$ and $X_g$ are the same.

$|x, x'|$ may be the metric of $X_g$. From the set of all bounded real continuous functions $\varphi : \mathbb{R} \to \mathbb{R}$ with compact support can be selected a countable subset $\{\varphi_\mu\}$ such that this subset is dense in the total set relative to the norm $\|\varphi\| = \sup_\alpha |\varphi(\alpha)|$ (see [10, Chap. X, Sect. 3, n.3]).

With a denumerable subset $\{x_\nu\} \subset X_g$, which is dense in $X_g$, are defined

$$f_{\nu\mu} = \varphi_\mu(|x, x_\nu|).$$

By these mappings $f_{\nu\mu} : X_g \to \mathbb{R}$ is then defined (proved above) an initial uniform structure $p$ which is identical to the initial uniform structure defined by *all* mappings of the form

$$f_\nu = \varphi(|x, x_\nu|),$$

and $p$ is metrizable and $X_p$ is precompact. It remains to show that $g$ is finer than $p$, and that the topologies of $X_p$ and $X_g$ are the same.

Since all mappings $f_\nu : X_g \to \mathbb{R}$ are uniformly continuous, the uniform structure $g$ is finer than $p$. Therefore the topology of $X_g$ is finer than the topology of $X_p$. The topologies are then the same if the identical mapping $X_p \to X_g$ is continuous. A vicinity $U$ of $x_0$ may be defined by $|x, x_0| < \varepsilon$; we seek a vicinity $V$ of $X_p$ with $V \subset U$. To this purpose we chose a $\varphi$ and a $\delta$, such that from $|\varphi(\alpha) - \varphi(0)| < \delta$ follows $|\alpha| < \varepsilon$. By $|\varphi(|x, x_0|) - \varphi(0)| < \delta$ is then determined in $X_p$ a vicinity $V$ of $x_0$ with $V \subset U$.

## 4.7 Do the "Laws of Nature" Describe Realities?

There was a time when one tried to base a physical theory on experimental facts. For instance it was said that Coulomb's law could be deduced from more and more experiments, in other words by conclusive induction. But all these considerations were not successful. It can only be stated that all (or at least practically all) of the experiments made until now are, within the scope of the inaccuracies, not in contradiction with this law. This is nothing other than what we consider as "the mathematical theory $MT_\Delta \overline{A}$ is without contradiction for all $\overline{A}$."

The question of the deduction of physical laws would then mean "Is it possible to give a method of deduction of $MT_\Delta$ from $MT_\Theta \overline{A}$ if $\overline{A}$ is large enough?" There are two major difficulties:

1. There is no possibility of deducing from some facts other facts (at least without using a physical theory; see Chap. 6).
2. There are for given $\overline{A}$ many possibilities to invent various $\Delta$ such that $MT_\Delta \overline{A}$ is without contradiction.

The first difficulty concerns not only the question of the deduction of physical laws from physical facts, but also the question: "Why do we trust our physical theories?" The only argument is that we find contentment by trusting in our physical theories. But without any physical theory one cannot say anything (starting with a given $\overline{A}$) about the next experiment. A *deduction* of a theory is impossible. Also, for the simplest case where we have written down in $\overline{A}$ many experiments of "dropping a stone from the tower of Pisa" and recording the fact that they all fell downwards, we cannot *deduce* that the next stone will also fall downwards. We can only construct a physical theory by introducing the mathematical axiom that "all" of the stones will fall down (here we have used the interpretation of mathematical symbols by (cor)). We can only *guess* the axioms and *trust* our theories.

## 4.7 Do the "Laws of Nature" Describe Realities?

An efficient question is thus the following: Is it possible to develop a method of guessing physical laws, and why do we trust physical laws?

Our answer to this questions is as follows:

If we have a very small application domain $A_p$ and a simple relation, it is sometimes not difficult to guess from experimental data a mathematical representation of this relation, e.g., if we consider the pressure $p$ and volume $v$ of a gas. If one has learned a little about mathematics, one will guess to represent this relation by "$pv = $ const." for constant temperature. But this guessing by only "looking" at the experimental result is impossible for greater application domains $A_p$.

If we look at the history of physics, many essential processes of guessing physical laws had a curious point of departure. One says that Newton was lying in his garden and saw an apple falling from a tree. This brought him to the idea that the moon is also falling toward the earth, but only with a much smaller acceleration. This would have led him to his gravitational theory.

More curiously was the invention of the law of "black radiation" by Planck. He tried to calculate this law in several ways. In one way he introduced for the energy of a harmonic oscillator a discrete series of values: $nh\nu$ ($n = 0, 1, 2, \ldots$ and $\nu = $ frequency of the oscillator). He introduced this series with the intention to let go later $h \to 0$. But he saw that a finite value of $h$ had led to a law which represented very well the experimental results. This was the starting point for the development of quantum mechanics!

It is interesting to read the papers of Einstein about his "general relativity theory." It is interesting to see his way of finding the "natural laws" of this theory. The way of finding is often not the way of understanding. Einstein called his theory the general relativity theory; nevertheless, his theory describes not a general relativity but the *absolute* metric structure of the four-dimensional differentiable manifold of space–time.

In a similar way, atoms were called atoms (i.e., indivisible objects); nevertheless, they are divisible.

Quantum mechanics was found in two different ways: the "matrix mechanics" of Heisenberg, Born, and Jordan and the "wave mechanics" of Schrödinger. But what is the correct interpretation of these two mathematical structures? One of the authors of this book has tried to formulate such an interpretation by an "axiomatic basis" $MT_{\widehat{\Sigma}}$ (in the form of Sect. 4.3) (see [8]).

Since we have often seen that the way of finding physical theories was not the way of formulating and interpreting these theories, the authors of this book are opposed in principle to all restrictions and to any prescriptions. "Everything" is permitted in the guessing of physical laws.

A physical theory becomes a serious part of science, not in the way of finding such a theory, but in the way of an exact formulation (in such a way as we have tried to describe this formulation in Chaps. 3 and 4). It becomes, in this way, a theory in which we can trust. But why?

At first it is necessary to believe that the reality, which we try to describe by a theory, is not an absolute chaos but a structured reality, that the reality does not have the aim of confusing us. If we speak today of chaos in physical processes, this "chaos" is structured not only in the way that the process is determined, but also by what has happened before. If the reality could confuse us in every respect, we would not be able to live. Without the belief in a structured reality no craft would be possible; and physics is nothing other than a further development of crafts by theories and techniques.

Our problem in trusting in a physical theory is the question of whether a particular invented theory is indeed an approximate description of a real structure or an error that we have made. Since a theory cannot be deduced from the reality, we can only compare the theory with the reality. This comparison is the one basis on which we can set our trust. But there can also be a second basis; the structure of the theory itself and the connection of this structure with the structures of already well-tried theories.

The comparison of a theory with the reality is made by the investigation of $MT_\Delta \overline{A}$ as outlined in Sect. 3.3.2. In Sect. 3.3.2 we have only described the principal method. But can we be convinced that a theory is correct if any test $\overline{A}$ is not in contradiction to $MT_\Delta$? Obviously not. If we take as $\overline{A}$ the measured distance between two marked spots, whereby one marked spot is used for the measurement of *one* distance, we cannot have a contradiction to $MT_\Delta$. This shows that these measurements are suitable for making a relevant test of the theory.

One has to try to find a "critical" experiment, i.e., an experiment by which an axiom or a theorem in $PT_\Delta$ can perhaps be refuted. In our Example A (applied to a round table) we can, e.g., perform the following experiment.

According to the pre-theory of measurements of distance, we use measuring tapes. We stretch a measuring tape between two marked spots $a$ and $b$; the length of this tape is a measured distance $\delta(a,b)$ of the marked spots $a, b$. (We do not use a ruler, since the definition of a ruler already presumes the Euclidean geometry; we also do not use compasses, since this presumes the definition of a "rigid body." We do not consider constructions with ruler and compasses, as Greek mathematicians have formulated in several mathematical problems.)

We begin the experiment with two marked spots $a, b$. We take a tape, the length of which is greater than the distance $\delta(a,b)$. We fix the ends of this tape onto $a$ and $b$. The tape is not stretched, since the length is greater than $\delta(a,b)$. But we can stretch the tape by joining the middle of the tape to a marked spot $c$, so that $\delta(a,c) = \delta(b,c) = \alpha$ with $2\alpha$ having the length of the tape. But there is also a second marked spot $d$, by which the tape is also stretched, so that $\delta(a,d) = \delta(b,d) = \alpha$ (see Fig. 4.3). We can now measure the distances $\delta(a,b)$ and $\delta(c,d)$. We define $\beta$ and $\gamma$ by $2\beta = \delta(a,b)$ and $2\gamma = \delta(c,d)$.

## 4.7 Do the "Laws of Nature" Describe Realities?

We can now ask whether these measured values $\alpha, \beta, \gamma$ fulfill with a certain inaccuracy the equation

$$\alpha^2 = \beta^2 + \gamma^2.$$

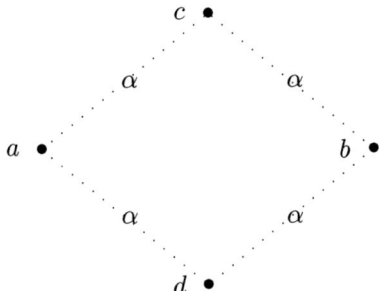

**Fig. 4.3.** Experiment with marked spots

If this is the case, we have no contradiction in $MT_\Delta \overline{A}$. If we repeat this experiment a few times, starting with other marked spots $a, b$, and get no contradiction to $MT_\Delta$, we are convinced that our theory is correct, i.e., we trust our theory. We trust it because we do not believe that the reality will fool us.

We can now repeat the same experiment on the surface of a globe (instead of the surface of the earth, since on earth it is more difficult to measure distances, i.e., since we need for the earth a richer theory of measurement of distances). The result of such experiments is then the following: the equation $\alpha^2 = \beta^2 + \gamma^2$ is only fulfilled with small inaccuracies if the $\alpha, \beta, \gamma$ are small compared to the greatest distance of two marked spots on the surface of the globe. We can either introduce large inaccuracies and as a consequence a smaller fundamental domain $G$ than the total globe, or to form a "better" theory by instead of using an Euclidean idealization, the idealization of the geometry of the surface of a sphere (see Sect. 3.3.2).

It is not possible to give a *general* method for finding and for making critical experiments; this depends on the selected theorems which we try to refute and on the expenditure of such an experiment.

But if we also have very rare contradictions in $MT_\Delta$ with $\overline{A}$ (contradictions that we cannot repeat systematically) then we do not refute our theory. We trust our theory because contradictions are very rare; – "practically" impossible.

It can also be that we trust a new theory even if we have made no critical experiments. For example, Einstein believed in his gravitational theory before any critical experiments, such as the deviation of light by the sun, were made; he knew that the experiments would confirm his theory! But why?

Before Einstein's new gravitational theory there were two other well-tried theories: the general relativity theory and Newton's gravitational theory. But it was not possible to unify these two theories. Einstein succeeded in achieving this in such a perfect way that his theory could not be wrong.

There is another point of view in which we can also judge a theory. We believe not only in a structure of the reality but also in the beauty and the intelligence of this structure.

In any case, we find contentment in developing physical theories. The essential problems today are not the physical theories, but the question of knowing what we *should* do and what we should not do on the basis of a physical theory.

## 4.8 Classification of Laws of Nature

In the preceding section we asked ourselves why we believe that a physical theory describes the reality in such a way that we can trust it. Now we want to ask what the various axioms of a simple axiomatic basis can tell us about the reality and, in particular, about the structure of "nature" itself. These considerations can also provide us with some hints for finding "critical experiments" as described in Sect. 4.7.

We will not speak about experiments which have already been made but only about *hypothetical experiments*. These hypothetical experiments can be defined purely mathematically. We want to study whether axioms of a simple axiomatic basis can be tested or perhaps deduced from hypothetical experiments.

A hypothetical experiment is defined by an additional text to $MT_{\widehat{\Delta}}$ which has the same form as $\overline{A}$. Such a text, denoted by $\mathcal{H}$, consists of relations of the form (see Sect. 3.2.3)

$$x_i \in \overline{M}, \quad (x_1, x_2, \ldots) \in \overline{s}, \quad (x_1, x_2, \ldots) \in \overline{s}'. \tag{4.8.1}$$

The $x_i$ are new constants in the theory $MT_{\widehat{\Delta}}\mathcal{H}$. It is clear that we do not use a text $\mathcal{H}$ (4.8.1) which has contradictions in itself. If we know that real errors of measurement (given by pre-theories) must be larger than the smallest intervals $J$ (given by pre-theories), then one can also replace (4.8.1) by relations of the form

$$x_i \in \overline{M}, \quad (x_1, x_2, \ldots, J) \cap \overline{s} \neq \emptyset, \quad (x_1, x_2, \ldots, J) \subset \overline{s}'. \tag{4.8.2}$$

Our aim is to test the structure $\widehat{\Sigma}$ by hypotheses $\mathcal{H}$ or, more precisely, to test the axiom $\widehat{P}$ of $\widehat{\Sigma}$ by possible $\mathcal{H}$. We want not only to test the total $\widehat{P}$, but also parts of $\widehat{P}$. Therefore we consider the following situations:

## 4.8 Classification of Laws of Nature

$\widehat{\Sigma}_0$ may be the text $\widehat{\Sigma}$ without the axiom $\widehat{P}$. $R_1$ may be a transportable relation in $MT_{\widehat{\Sigma}_0}$ which is a theorem in $MT_{\widehat{\Sigma}}$, e.g., a part of the axiom $\widehat{P}$.
$\widehat{\Sigma}_1$ may be the text $\widehat{\Sigma}_0$ plus the relation $R_1$. Since $R_1$ is a theorem in $MT_{\widehat{\Sigma}}$, the species of structures $\widehat{\Sigma}_1$ is said to be poorer than $\widehat{\Sigma}$. (If we introduce no $R_1$, then $\widehat{\Sigma}_1 \equiv \widehat{\Sigma}_0$.) $R_2$ may be a transportable relation in $MT_{\widehat{\Sigma}_1}$ which is a theorem in $MT_{\widehat{\Sigma}}$, e.g., a part of the axiom $\widehat{P}$.
$\widehat{\Sigma}_2$ may be the text $\widehat{\Sigma}_1$ plus the relation $R_2$ (i.e., $\widehat{\Sigma}_0$ plus the relation "$R_1$ and $R_2$"). Since "$R_1$ and $R_2$" is a theorem in $MT_{\widehat{\Sigma}}$, the species of structures $\widehat{\Sigma}_2$ is said to be poorer than $\widehat{\Sigma}$, but richer than $\widehat{\Sigma}_1$.

We have the chain $\widehat{\Sigma}_0, \widehat{\Sigma}_1, \widehat{\Sigma}_2, \widehat{\Sigma}$, where (from the left to the right) the next species of structures is richer than the one before.

Now we consider the chain of species of structures

$$\widehat{\Delta}_0, \widehat{\Delta}_1, \widehat{\Delta}_2, \widehat{\Delta}$$

generated by the chain $\widehat{\Sigma}_0, \widehat{\Sigma}_1, \widehat{\Sigma}_2, \widehat{\Sigma}$ in the way described in Sect. 4.3. From the form (4.3.2) for the axioms of the $\Delta$ we see immediately that in this chain the next species of structures is richer than the previous one.

We introduce the following classification of $R_2$ relative to $R_1$:

We consider only such hypotheses $\mathcal{H}$ for which $MT_{\widehat{\Delta}_1}\mathcal{H}$ contains no contradictions. We call such a hypothesis $\mathcal{H}$ a "$R_1$-allowed" hypothesis. We call a relation $R_2$ an "empirically allowed relation relative to $R_1$" if $MT_{\widehat{\Delta}_2}\mathcal{H}$ contains no contradictions for all "$R_1$-allowed" hypotheses $\mathcal{H}$. We call a relation $R_2$ an "empirically deducible relation relative to $R_1$" if there is a "$R_1$-allowed" hypothesis $\mathcal{H}$, so that the axiom of $\widehat{\Delta}_2$ is a theorem in $MT_{\widehat{\Delta}_1}\mathcal{H}$. We call a relation $R_2$ an "empirically refutable relation relative to $R_1$" if there is a "$R_1$-allowed" hypothesis $\mathcal{H}$, so that we get a contradiction in $MT_{\widehat{\Delta}_2}\mathcal{H}$. $R_2$ not "empirically allowed" is equivalent to the fact that there is a hypothesis $\mathcal{H}$ such that $MT_{\widehat{\Delta}_2}\mathcal{H}$ contains a contradiction, i.e., $R_2$ is empirically refutable.

What is the physical significance of this classification?

(1) $R_2$ is an empirically allowed relation relative to $R_1$.

In this case the axiom $R_2$ has no influence on the assertion of the physical theory about the reality. What was allowed by the theory with the axiom $R_1$ is also allowed with the axiom "$R_1$ and $R_2$." We therefore call $R_2$ a "pure idealization." One can add $R_2$, e.g., to simplify the mathematical structure of $\widehat{\Sigma}$. The fact that $R_2$ is empirically allowed depends on the used inaccuracy sets. We say that $R_2$ is an "absolutely pure idealization" if it is empirically allowed for *all* elements of the uniform structures taken from the reservoir of inaccuracy sets (see Sect. 4.6). Such an absolutely pure idealization was, e.g., introduced as an axiom AVid for an axiomatic basis for quantum mechanics (see [8] III, Sect. 3). Another example can be found in our Example B (see Part II).

(2) $R_2$ is an empirically refutable relation relative to $R_1$.

In this case the axiom $R_2$ *restricts* the hypotheses $\mathcal{H}$, which is what we want! Any hypothesis $\mathcal{H}$ which produces contradictions with $\widehat{\Delta}_2$ is not possible as an $\overline{A}$ describing a reality. The aim of the "physical laws" is to restrict the hypotheses to a hypothesis $\mathcal{H}$ which can be "real." But this aim is more than the condition that $MT_{\widehat{\Delta}}\overline{A}$ produces no contradiction for the "real" $\overline{A}$. There could also be a hypothesis $\mathcal{H}$ which produces no contradiction but cannot be real! The aim of a physical theory is therefore more than only the condition of "not being in contradiction to the reality" of Popper.

(3) $R_2$ is an empirically deductible relation relative to $R_1$.

This case is very rare. We have already discussed this problem in Sect. 4.7 of whether it is possible to deduce a physical law from experiments, i.e., from a suitable $\widehat{\Delta}$. The reason that this case is rare is that $R_2$ contains in most cases the universal quantifier $\forall$, while a hypothesis $\mathcal{H}$ contains only finite many $x_i$. Also if a set is finite, one never knows how many elements this set has or if the $x_i$ are "all" the elements!

The only interesting case is therefore the case where $R_2$ is refutable relative to $R_1$. In this case there is an $R_1$-allowed hypothesis $\mathcal{H}$ for which $MT_{\widehat{\Delta}_2}\mathcal{H}$ contains a contradiction! Why do we not consider the case where we have "no $R_1$," i.e., where all hypotheses $\mathcal{H}$ are allowed? The reason is that in practically all theories we do not allow that there are contradictions between the reality and some axioms (laws) of $\widehat{\Sigma}$. Therefore we incorporate in $R_1$ all such "laws" where we do not allow a contradiction to the reality. We call such laws *norms*. There are several reasons for introducing norms:

(a) Conceptual reasons: We use in the basic language a concept where this concept includes a structure, the negation of which would contradict the concept. For example, the concept "part of," used in sentences such as 'the leg of a chair is part of the chair.' This concept includes the "law," that from "$a$ is a part of $b$" and "$b$ is a part of $c$" follows that "$a$ is a part of $c$." Therefore we introduce in $\widehat{\Sigma}$, where the picture of this relation may be symbolized by the sign "$<$," the axioms "$a < b$ and $b < c \Rightarrow a < c$." If we would have written in $\overline{A}$ a relation contradicting this axiom, we would not say that the reality contradicts the theory; we would say that we have made a mistake by the application of the concept "part of." Therefore we incorporate the law "$a < b$ and $b < c \Rightarrow a < c$" in $R_1$. This "physical law" is not a "law of untouched nature" but a "law of a (physical) concept."

(b) The reality that we want to describe by a physical theory is not only the "untouched nature" (e.g., a star) but also contains constructions made by human beings (e.g., a car). There is no physics without craft and modern techniques. The structure of such constructions is not only determined by laws of nature, but also decisively by a plan, i.e., by a "required structure." These required structures appear in $\widehat{\Sigma}$ also as axioms, which we call "norms."

The first great physical theory, Newton's mechanics, already contains such norms. One part of this theory is a space–time theory. How has Newton introduced space and time?

He started with the introduction of an "absolute space" and an "absolute time." The absolute space is introduced as an Euclidean three-dimensional space, the absolute time as a one-dimensional manifold with a transportable distance of time points. Time can be represented by the set $\mathbb{R}$ (of real numbers) with $|\alpha - \beta|$ as the distance between the times $\alpha$ and $\beta$.

This introduction of the absolute space and absolute time is nothing other than what we have called a "fairy tale." This fairy tale becomes a physical theory only by the introduction of terms which can be interpreted by "real facts," i.e., by terms which are connected to recordable facts by (cor) and the mappings $\phi$ between $MT_\Theta$ and $MT_\Sigma$. These terms were introduced as "space–time reference systems."

Before we speak about the reality which will be compared to these mathematically introduced reference systems, we will include some remarks concerning the problem of the reality of an absolute space and of an absolute time. It may be that Newton believed in these realities. In any case there were many physicists (practically all) who believed in these realities before the introduction of the special relativity theory. The belief was so strong that there were many attempts to disprove the special relativity theory and Einstein's gravitational theory. Nevertheless, there are practically no physicists today who believe in the reality of an absolute space and of an absolute time. But why? This will be one of the problems dealt with in Chap. 6.

What are the realities from which reference systems which have been mathematically introduced are pictures? Such real reference systems cannot be found in untouched nature. They are constructions made by human beings:

(a) The mathematical time scale will be the picture of the real scale of "clocks." The construction plan of a "clock" is (very briefly speaking) to repeat the same process and to count the number of these processes. The main purpose of a clock is the "repetition of the same processes." The best realization of the repetition of the "same processes" was obtained by the "clock in Braunschweig;" this clock is better than the repetition of the rotations of the earth.

There are, however, many clocks. In order to obtain "one" time, one has to fulfill two other norms: the norm of the same unit for all clocks and the norm of a synchronization of all of these clocks. Only if these norms are fulfilled can we have a real situation which we can mathematically describe by one time $t$, as in classical physics. The main problem is the synchronization, i.e., the norm which ensures that every pair of two clocks (and therefore all clocks) indicate the same time when we compare the two clocks together. We know today that this norm can only be approximatively fulfilled by a set of clocks which do not move too fast one in relation to the other, and if there is no great gravitational field. The "one time" in classical mechanics is a norm. Only if this norm is (approximatively) fulfilled, can we "apply" this classical mechanics. Thus we get by the "norms" a restriction of the application domain $A_p$, that is the fundamental domain $G$ (see Sect. 3.3.3).

We wish to insert here some remarks concerning a concept sometimes used in the description of physics even though this concept is *not* a physical concept. We mean the concept "now." We say: it is 'now $t$-a clock'. This clause cannot be transcribed by (cor) in a relation in $MT_\Theta$ because there is no physical theory. By the word 'now' we designate a concept belonging to a subjective experience in our consciousness. We call this 'now' also the 'present'. Therefore words like 'tomorrow' and 'yesterday' do not also appear in physical theories. They can appear, e.g., if a physicist speaks of his work in a laboratory. There is in physics no possibility to demonstrate the reality of what "is" real, what "was" real, and what "can be" real in the future. The word "real" in physical theories can only be related to a "recording of a time $t$ on a clock." In physical theories, there are no parts of sentences with the words 'was real' or 'will be real'. It can be that such words describe only a subjective experience in our consciousness.

(b) The space of a real reference system is also a construction determined by norms.

The first aim of these norms is to obtain a "rigid basis." If we could swim like a fish in the ocean, it would be impossible to have a reference system. We would not know how to construct it. We could swim to the ground in order to try to construct, starting from this ground, a reference space for the description of the movement of the water. There is no absolute space that one could take as a reference space. It could be an interesting question to ask! Is it possible to develop physics without rigid bodies? And if so, how?

There are many ways in which to construct reference spaces. For instance, the office or laboratory where we work. But also the reference space that Kepler has constructed to formulate his laws for the motion of the planets. But where is the rigid part of this reference space? It is found by the sun and the fixed stars connected by rays of light.

There are several systems of construction of a mathematical picture given for a space reference system (see [11–13]). In these two cases the norms of construction are such that we obtain an Euclidean space. It is not difficult to see that these norms cannot be realized for arbitrary high distances. We know that it is impossible to realize such a space reference system for all the stars in the universe; that is, the total universe does not belong to the fundamental domain $G$ of this theory.

The given examples will suffice to explain the following classification of physical laws: We want to distinguish between "norms" and "laws of untouched nature." The norms are taken together into the relation $R_1$, the other axioms into the relation $R_2$. It would be a misunderstanding to say that the "norms" have nothing to do with the structure of nature. Norms, when there is no region where it is possible to realize them, are without significance as physical laws; that is, the norms make visible structures of nature by the fact that they can be realized.

In separating the norms $R_1$ from the other laws $R_2$, we can restrict the application domain $A_p$ to the fundamental domain $G$ in two steps: the first step consists in using only $R_1$, the second step consists in using "$R_1$ and $R_2$." To do this, we can use the method explained in Sect. 3.3.3. We also use for $R_1$ inaccuracy sets so that we get no contradiction to reality, and then exclude from the application domain $A_p$ the realities where the inaccuracies are very large. Thus we get a fundamental domain $G_1$ where the norms $R_1$ can be realized with small inaccuracies. If the total fundamental domain $G$ is smaller than $G_1$, then we can try to improve $R_2$, if possible, in such a way that the fundamental domain $G$ is equal to $G_1$.

At the end of this section we would like to say something about the introduction of theoretical terms. As terms of $MT_{\widehat{\Sigma}}$ with physical significance, we have only introduced the picture terms $\widehat{M}_i$ and $\widehat{s}_\nu$ as pictures of the terms $\langle \overline{M}_i, \overline{s}_\nu \rangle$ of $\Theta$. The physical significance of the $\langle \overline{M}_i, \overline{s}_\nu \rangle$ were given by (cor). After the introduction of $MT_\Sigma$ and $MT_{\widehat{\Sigma}}$, one has often denoted many other terms with words, which sound physical. We will not introduce such *general* theoretical terms. We let terms in $MT_\Sigma$ and $MT_{\widehat{\Sigma}}$ be only mathematical terms. We will later give to some of these terms a realistic significance (see Chap. 6).

## 4.9 Skeleton and Uninterpreted Theories

At the end of the description of the method of a physical theory we will mention a very useful mathematical method for dealing with $MT_\Sigma$ and $MT_{\widehat{\Sigma}}$. A pure mathematical theory $MT_\Sigma$ (or $MT_{\widehat{\Sigma}}$) becomes a part of a physical theory by the transition from $MT_\Theta$ to $MT_\Delta$ (or $MT_{\widehat{\Delta}}$). The species of structures $\Theta$ has by (cor) a physical interpretation. There is, in general, a lot of work to be done to elaborate the structure of the $MT_\Sigma$. For this we do not need the connection of $MT_\Sigma$ with $MT_\Theta$ by $MT_\Delta$. Moreover, it appears that for different $MT_{\Theta_1}$ and $MT_{\Theta_2}$ we can use the same $MT_\Sigma$.

Therefore, one often develops the structure of a $MT_\Sigma$ using more or less precisely physical words without introducing an exact definition of these words by a $\Theta$. But it can be that some of the terms of $MT_\Sigma$ have a precise interpretation, and others not. We will call such a $MT_\Sigma$ a skeleton theory.

The very well elaborated $MT_\Sigma$ of the classical mechanics of mass points is such a skeleton theory.

The basic structure of $MT_\Sigma$ are the equations of motion for the $\vec{r}_i(t)$:

$$m_i \ddot{\vec{r}} = \vec{f}_i(\vec{r}_1, \vec{r}_2, \ldots, \vec{r}_n).$$

Here, $t$ is the time exactly interpreted and the $\vec{r}_i$ are the positions exactly interpreted in a space–time reference system, but *without* saying in which. The $\vec{f}_i$ are functions which will be *later* interpreted as "forces." But firstly one plays with these functions introducing several conditions (i.e., mathematical axioms) for these $\vec{f}_i$. The indices $i$ are interpreted as signs for several "mass points" *without* saying whether and where these mass points are in reality. Mathematicians such as Hamilton and Jacobi have elaborated this skeleton theory in an admirable way.

Another skeleton theory can be found in thermodynamics. One considers an $n$-dimensional state space $(x_1, \ldots, x_n)$ without giving a physical interpretation of these $x_\nu$, but with the intention that they will be interpreted later. One introduces the differential 1-form

$$\mathrm{d}A = \sum_{\nu=1}^{n} f_\nu(x_1, \ldots)\, \mathrm{d}x_\nu,$$

with the interpretation that $\mathrm{d}A$ is the picture term for the work which is added to the systems if the states are changed by $\mathrm{d}x_\nu$. The axiom of this skeleton theory is: there are three functions $U(x_1, \ldots)$, $T(x_1, \ldots)$, and $S(x_1, \ldots)$ with the relation

$$\mathrm{d}U + \mathrm{d}A = T\mathrm{d}S.$$

One calls $U$ the intrinsic energy, $T$ the temperature, and $S$ the entropy.

The application of this skeleton theory is done by introducing exactly interpreted states $(x_1, \ldots)$ and a $\mathrm{d}A$ given by pre-theories.

This method of skeleton theories, briefly described, is in total conformity with our foundation of the method of a physical theory, since the noninterpreted terms are introduced with the intention that they will later be specialized and interpreted.

But there are also descriptions of physical theories which contradict the method described above. One introduces a fairy tale theory *without* an interpretation in our sense (i.e., an interpretation as described in Sect. 4.3). This

## 4.9 Skeleton and Uninterpreted Theories

means that the connection of mathematical terms with physically sounding words is enough to obtain a connection between the theory with experimental results. One has in these theories only an imagined reality without knowing what this imagined reality has to do with the true reality. Very often we introduce also a priori conditions for the structure of these fairy tales, because one believes that reality *must* have these structures.

There is for instance the belief that physics should be founded on quantum theory, i.e., that one can begin with the description of an intrinsic reality by quantum theory, and that one can get, from this intrinsic reality, the reality of a stone that we can see and feel without knowing any physical theory. One can even find that the meaning of this intrinsic reality is different from (and perhaps in contradiction to) the subjective reality of a stone that we perceive with our senses. The authors of this book believe that the reality of the stone that we can see is the same with or without physics. We learn only from physics that there are also other realities as, e.g., electrons, protons, etc. But an electron can be recognized only if we have already recognized, without physics, the reality of a stone.

In this sense is described the quantum theory in [8] where a theory of preparation and registration procedures, which is physically interpreted without quantum mechanics (see [8, Chaps. II and III, Sects. 1–4]), is introduced as a basis of a quantum theory. The axiom AQ in Chap. III, Sect. 5 is then the connection of these interpreted terms with terms of the Hilbert space theory. Thus we get an axiomatic basis for the quantum theory, but, because of the axiom AQ, not a simple one. A simple axiomatic basis is given in [8]. The formulation and interpretation of this axiomatic basis is founded on the reality of macroscopic events, the preparation and registration procedures, a reality known before all quantum mechanics.

In contradiction to this description of quantum mechanics there are pure fairy tales, the reality of which is more than dubious. One tries to base the theory on "states" of the objects represented by vectors of a Hilbert space without knowing what the reality of such "states" is. One introduces a "superposition principle" without saying what the reality of such a "superposition" is (one cannot superpose two microsystems in order to get a new microsystem in such a way that one can superpose two waves of water in order to get a new wave of water). All this would not contradict our method of a physical theory if one could formulate this fairy tale in a correct mathematical form $MT_\Sigma$, *and* if one could define in this $MT_\Sigma$ the picture terms for a reality which is known before quantum mechanics and can be described by a $MT_\Theta$. We have never seen such a formulation.

Firstly, we have not seen a correct mathematical form $MT_\Sigma$. For instance one formulates a sentence of the form 'the microsystem $a$ has at the instant $t$ the state $\varphi$' (or '... is in the state $\varphi$ ...'), where $\varphi$ is a vector of a Hilbert space and $t$ a real number. But such a sentence is not a mathematical one. In Sect. 3.1 we have formulated what mathematical sentences should look like. The above sentence can indeed be *transcribed* into a formal sentence or a mathematical form. For example, we introduce a relational sign of weight 3 for this relation between $a, t, \varphi$ and write for the above natural sentence $r(a, t, \varphi)$. Then we can introduce two sets, a set $M$, the elements of which shall be the picture terms of the microsystems, and a set $H$ of vectors of the Hilbert space. With $a \in M$, $t \in \mathbb{R}$, and $\varphi \in H$, the relation $r(a, t, \varphi)$ can be replaced by a subset $s \subset M \times \mathbb{R} \times H$. Then one can introduce the axiom that $s$ defines a mapping $g : M \times \mathbb{R} \to H$. The above natural sentence can then be transcribed into the mathematical form $g(a, t) = \varphi$. In this way one could perhaps formulate a correct mathematical form $MT_\Sigma$ of this fairy tale.

Secondly, we have not seen a correct formulation for the terms (in $MT_\Sigma$) which can be compared to the results of experiments, e.g., for preparation and registration procedures.

Until now this theory is not a correct formulation $MT_\Sigma$ of a fairy tale, as required in Sect. 4.3.

Instead of such a formulation $MT_\Sigma$, one finds in this theory very incredible fairy tales such as the collapse of wave packets (i.e., of states) by measurements. Since this appeared too mysterious, one tried to interpret the vectors of the Hilbert space as a description of "the knowledge of a subject." But such a way out is not necessary since there is no collapse of wave packets in reality. Do not believe in fairy tales! Consider only the reality of fairy tales in the form of our investigations in Chap. 6.

# 5. Relations Between Various $PT$s

In Chap. 4 we finished formulating the method of one physical theory. It is only by following this method that we accept a set of sentences (using physical and mathematical concepts) as a $PT$. This does not mean that we do not accept a $PT$ if all the steps are not correctly followed according to our formulated method. Also, the mathematicians do not need to elaborate all of the proofs; it suffices that we see only the main steps. The physicists use mathematics with still greater negligence than the mathematicians, hoping that mistakes will be detected also by the applications of $MT$ in $PT$. But if there is a problem, whether one has made mistakes or not, one can return to the "correct" method.

But physics consists not only of *one* theory; it is made up of a set of various theories. Most physicists think that there should be, at least as an aim (which can perhaps never be reached), a single theory for all of physics. But this is not the case in physics today. We have many theories, but not only isolated ones. Sometimes one theory is "better" than another. But by "better" what does this mean? One is interested not only in better and better theories, but also in the "not so good" theories (so-called "approximation theories") which are mathematically simpler to elaborate and therefore to apply than the better theories.

The first aim of this chapter is to formulate more correctly what we mean by $\lceil PT_\alpha$ is a "better" theory than $PT_\beta \rceil$.

The second aim is to formulate the relation that a $PT_\alpha$ is used as a "pre-theory" for $PT_\beta$. This relation was already used in the formulation of the basic language $B_l$ of the theory, i.e., by the introduction of some of the concepts used in this language and in the formulation of recorded facts in the basic language (see Sect. 3.1.4).

## 5.1 Relations Between Two *PT*s with the Same Application Domain

We can symbolize the formulation of one theory $PT_\alpha$ under the form

$$PT_\alpha \equiv A_{p_\alpha} \leftrightarrow B_{l_\alpha}(\text{cor}) MT_{\Delta_\alpha}$$

in which "$\leftrightarrow B_{l_\alpha}(\text{cor})$" is the process by which we have connected the mathematical theory $MT_{\Delta_\alpha}$ with that part of reality which we called the application domain $A_{p_\alpha}$. The application domain $A_{p_\alpha}$ is that part of reality which can be described by the concepts used in the basic language $B_{l_\alpha}$ and which only contains such objects which have at least one of the basic properties (see Sect. 3.1.1.2).

We want to consider at first two theories $PT_\alpha$ and $PT_\beta$ with the same application domain $A_{p_\alpha} \equiv A_{p_\beta}$. Later we will try to extend these considerations to more general cases.

$A_{p_\alpha} \equiv A_{p_\beta}$ is equivalent to the fact that the relations $\widetilde{A}$ of $PT_\alpha$ are also the relations of $PT_\beta$ and vice versa. This is equivalent to the fact that $B_{l_\alpha} \equiv B_{l_\beta}$, and that the basic properties are the same. Only $\Delta_\alpha$ and $\Delta_\beta$ can be different. Also the parts $\Theta_\alpha$ of $\Delta_\alpha$ and $\Theta_\beta$ of $\Delta_\beta$ must be the same, i.e., the base sets and the structure terms of $\Delta_\alpha$ and $\Delta_\beta$ are the same. Only the axiomatic relations $P_{\Delta_\alpha}$ of $\Delta_\alpha$ and $P_{\Delta_\beta}$ of $\Delta_\beta$ can be different (for the axiomatic relation, see Sect. 4.1).

The axiomatic relation $P_{\Delta_\alpha}$ of $\Delta_\alpha$ is determined by the idealization $\Sigma_\alpha$ and the used inaccuracy sets. We use for $\Sigma_\alpha$ the form of an axiomatic basis (for the axiomatic basis, see Sect. 4.3). Since $\Theta_\alpha \equiv \Theta_\beta$ we get that the base sets and the structure terms of $\Delta_\alpha$ and $\Delta_\beta$ are the same; only the axiomatic relations $P_\alpha$ of $\Sigma_\alpha$ and $P_\beta$ of $\Sigma_\beta$ can be different. Since the base sets and the structure terms of $\Sigma_\alpha$ and $\Sigma_\beta$ are the same, we also take the same unitary structures for the "reservoir" of inaccuracy sets. The difference between $P_{\Delta_\alpha}$ and $P_{\Delta_\beta}$ can only be generated by different inaccuracy sets, selected in the common unitary structure, and by different axiomatic relations $P_\alpha$ and $P_\beta$ of the axiomatic basis $\Sigma_\alpha$ and $\Sigma_\beta$.

If $\Sigma_\alpha$ and $\Sigma_\beta$ are "equally rich" (see Sect. 4.1), which we write as $\Sigma_\alpha \equiv \Sigma_\beta$, then the difference between $\Sigma_\alpha$ and $\Sigma_\beta$ is only the *form* of the axiomatic relations $P_\alpha$ and $P_\beta$, which does not interest us here. For $\Sigma_\alpha \equiv \Sigma_\beta$ we also say that the idealizations of $\Delta_\alpha$ and $\Delta_\beta$ are the same. The differences between $\Delta_\alpha$ and $\Delta_\beta$ can then be generated only by different inaccuracy sets. If the inaccuracy sets of $\Delta_\alpha$ and $\Delta_\beta$ are also the same, it follows that $\Delta_\alpha$ and $\Delta_\beta$ are "equally rich," written $\Delta_\alpha \equiv \Delta_\beta$. Then $PT_\alpha$ is also equivalent to $PT_\beta$, written $PT_\alpha \equiv PT_\beta$.

Here we are no longer interested in equivalent physical theories (such as in Sect. 4.2 where we investigated different forms of the same physical theories). We are interested in relations of the form

⌈ $PT_\beta$ is "better" than $PT_\alpha$ ⌉,

that is, $PT_\beta$ says more about the structure of the application domain $A_{p_\alpha} \equiv A_{p_\beta}$. This is the case if ⌈ $\Delta_\beta$ is richer than $\Delta_\alpha$ ⌉. We always presume that we have no contradiction with experiment. Since $\Theta_\alpha \equiv \Theta_\beta$, the texts $\overline{A}_\alpha$ and $\overline{A}_\beta$ are equivalent, which we write as $\overline{A}_\alpha \equiv \overline{A}_\beta$. We only consider $PT_\alpha$ and $PT_\beta$ for which $MT_{\Delta_\alpha}\overline{A}_\alpha$ and $MT_{\Delta_\beta}\overline{A}_\beta$ are without contradiction for all experiments, and when we are convinced that they will also be without contradiction for future experiments (for this "conviction," see Sect. 5.4).

We define

⌈ $PT_\beta$ is "richer in content" (briefly "richer") than $PT_\alpha$ ⌉
if ⌈ $\Delta_\beta$ is richer than $\Delta_\alpha$ ⌉.

We write this in the form $PT_\beta \succ PT_\alpha$.

(A remark concerning the ideas developed in [1]. There, we had used for "richer in content" the term "more comprehensive" (German word: *umfangreicher*). In a discussion through letters, Prof. Erhard Scheibe has remarked that the word *umfangreicher* can be misunderstood. For instance, a concept $c_1$ contains more than a concept $c_2$, e.g., $c_1$ = mammal and $c_2$ = dog. Dr. Scheibe proposed to use the term "richer in content" (German word: *inhaltsreicher*) and we agreed with this proposal.

We see immediately that for $\Sigma_\alpha \equiv \Sigma_\beta$ that ⌈ $PT_\beta$ is richer than $PT_\alpha$ ⌉ if the inaccuracy sets for $\Delta_\beta$ are smaller than those of $\Delta_\alpha$. Therefore one will try to choose, for given $\Sigma_\alpha \equiv \Sigma_\beta$, inaccuracy sets as small as possible without contradictions to experiments, i.e., to $\overline{A}_\alpha \equiv \overline{A}_\beta$ for "all" experiments. Therefore, we are mainly interested in the case where $\Sigma_\alpha \not\equiv \Sigma_\beta$.

If ⌈ $\Sigma_\beta$ is richer than $\Sigma_\alpha$ ⌉ and $\Sigma_\alpha \not\equiv \Sigma_\beta$, we see immediately that ⌈ $\Delta_\beta$ is richer than $\Delta_\alpha$ ⌉ if we use the same inaccuracy sets for the construction of $\Delta_\alpha$ and $\Delta_\beta$. Thus ⌈ $PT_\beta$ is richer than $PT_\alpha$ ⌉ if $MT_{\Delta_\beta}\overline{A}_\beta$ is without contradiction like $MT_{\Delta_\alpha}\overline{A}_\alpha$: the experiments must "decide" whether we have indeed found in $PT_\beta$ a richer theory.

An example for this case is the following: The axiomatic relation $P_\alpha$ of $\Sigma_\alpha$ has the form $R_1 \Rightarrow R_2$ as described in Sect. 2.4. The axiomatic relation $P_\beta$ is a generalization of $P_\alpha$ in the sense that the "condition" $R_1$ is canceled, so that $R_1 \Rightarrow R_2$ is a theorem in $MT_{\Sigma_\beta}$. The axiomatic relation $P_\beta$ gives for the condition $R_1$ the particular relation $R_2$. ⌈ $MT_\beta$ is richer than $MT_\alpha$ ⌉ since it also describes structures of reality if the condition $R_1$ is not fulfilled.

Most interesting is another case, where neither ⌈ $\Sigma_\alpha$ is richer than $\Sigma_\beta$ ⌉ nor ⌈ $\Sigma_\beta$ is richer than $\Sigma_\alpha$ ⌉; the two structures $\Sigma_\alpha$ and $\Sigma_\beta$ contradict each other.

It may also seem that in this case $\Delta_\alpha$ contradicts $\Delta_\beta$, but this must not be the case since we may have used different inaccuracy sets for $\Delta_\alpha$ and $\Delta_\beta$! It can be that ⌈ $\Sigma_\beta$ is a "better" idealization than $\Sigma_\alpha$ ⌉, i.e., the inaccuracy of

$\Sigma_\beta$ can be smaller than that of $\Sigma_\alpha$ when compared to reality. Such cases are very interesting for physics, since one does not have in all interesting physical theories a $\Sigma_\beta$ which is a realistic idealization. Therefore we will concentrate on this case and will try to formulate this "better" by mathematical relations. We will provide a *method* for constructing usable inaccuracy sets which will allow us to compare $\Sigma_\alpha$ and $\Sigma_\beta$.

We will call the inaccuracy sets used before this construction the initial sets $U^{\alpha i}$ and $U^{\beta i}$. These sets are selected as small as possible under the condition that the so-constructed $\Delta_\alpha^i$ and $\Delta_\beta^i$ are without contradiction to $\overline{A}_\alpha \equiv \overline{A}_\beta$. It can be that neither $\lceil \Delta_\alpha^i$ is richer than $\Delta_\beta^i \rceil$ nor $\lceil \Delta_\beta^i$ is richer than $\Delta_\alpha^i \rceil$. Then we seek to construct such inaccuracy sets $U^{\alpha i}$ and $U^{\beta i}$ for which $\Delta_\alpha^f$ and $\Delta_\beta^f$ can be compared, i.e., for which $\lceil \Delta_\alpha^f$ is richer than $\Delta_\beta^f \rceil$ or $\lceil \Delta_\beta^f$ is richer than $\Delta_\alpha^f \rceil$. We postulate that the $U^{\alpha f}, U^{\beta f}$ are larger than the $U^{\alpha i}, U^{\beta i}$, so that the $\Delta_\alpha^f, \Delta_\beta^f$ are also without contradiction to $\overline{A}_\alpha \equiv \overline{A}_\beta$.

We begin with a question concerning $\Sigma_\alpha$ and $\Sigma_\beta$. The principal base sets $M_i$ are the same, the structure terms $s_{\nu\alpha}$ and $s_{\nu\beta}$ are different, the $s_{\nu\alpha}$ fulfill the axiomatic relation $P_\alpha$, and the $s_{\nu\beta}$ fulfill the axiomatic relation $P_\beta$. We consider the relations

$$s_{\nu\beta} \subset (s_{\nu\alpha})_{\widehat{U}_\nu} \quad \text{and} \quad s'_{\nu\beta} \subset (s'_{\nu\alpha})_{\widehat{U}_\nu}. \tag{5.1.1}$$

We try to find inaccuracy sets $\widehat{U}_\nu$ (as elements of the reservoir (see Sect. 4.6)) such that (5.1.1) is a theorem. We seek such $\widehat{U}_\nu$ which are as small as possible.

To obtain from $\Sigma_\beta$ the structure $\Delta_\beta^i$, the inaccuracy sets may be $U_\nu^{\beta i}$, i.e., we have as an axiomatic relation of $\Delta_\beta^i$ (see Sect. 3.3.5)

$$(\exists \phi_1)(\exists \phi_2) \cdots$$
$$[\phi_i : \overline{M}_i \to M_i \text{ are injective mappings}$$
$$\wedge \phi \overline{s}_\nu \subset (s_{\nu\beta})_{U_\nu^{\beta i}}$$
$$\wedge \phi \overline{s}'_\nu \subset (s'_{\nu\beta})_{U_\nu^{\beta i}}]. \tag{5.1.2}$$

From (5.1.1) and (5.1.2) follows the theorem

$$(\exists \phi_1)(\exists \phi_2) \cdots$$
$$[\phi_i : \overline{M}_i \to M_i \text{ are injective mappings}$$
$$\wedge \phi \overline{s}_\nu \subset \left((s_{\nu\beta})_{\widehat{U}_\nu}\right)_{U_\nu^{\beta i}}$$
$$\wedge \phi \overline{s}'_\nu \subset \left((s'_{\nu\beta})_{\widehat{U}_\nu}\right)_{U_\nu^{\beta i}}]. \tag{5.1.3}$$

We see that

$$(X_{\widehat{U}})_U = X_{\widehat{U} \circ U} \tag{5.1.4}$$

with $\widehat{U} \circ U$ according to Sect. 4.6.

Thus we get from (5.1.3) that the structure $\Delta_\alpha^f$, generated by $\Sigma_\alpha$ and the inaccuracy sets $U^{\alpha f} = \widehat{U}_\nu \circ U_\nu^{\beta i}$, is poorer than $\Delta_\beta^i$, i.e., $\lceil PT_\beta^i$ is richer than $PT_\alpha^f \rceil$.

This theorem does not contain any relation that specified that $\Sigma_\beta$ resembles more than $\Sigma_\alpha$ the structure of reality. A relation similar to (5.1.1) (with only $\alpha$ and $\beta$ permuted) can be proved for an $\widehat{U}$ which is as large as $\widehat{U}$ of (5.1.1). Thus it may seem that it is arbitrary to declare $\Sigma_\alpha$ or $\Sigma_\beta$ as resembling more the structure of reality. But this would be a deception. The essential difference lies in the formula (5.1.4). Starting from $\Sigma_\beta$ with $U^{\beta i}$ we get $U^{\alpha f} = \widehat{U} \circ U^{\beta i}$; starting from $\Sigma_\alpha$ with $U^{\alpha i}$ we get $U^{\beta f} = \widehat{U}' \circ U^{\alpha i}$ with an $\widehat{U}'$ which is as large as $\widehat{U}$. If $U^{\beta i}$ is much smaller than $U^{\alpha i}$, then $U^{\alpha f}$ is as large as $\widehat{U}$ and $U^{\alpha i}$, and $U^{\beta f}$ is much larger than $U^{\alpha i}$. If $U^{\beta i}$ is very small compared to $U^{\alpha i}$, then $U^{\beta f}$ is much larger than $U^{\beta i}$. To get $\lceil \Delta_\beta^f$ is richer than $\Delta_\alpha^i \rceil$, we have to change radically the inaccuracy from $U^{\beta i}$ to a much larger $U^{\beta f}$! This makes physically no sense, since $U^{\beta f}$ is *much* larger than necessary, as only $U^{\beta i}$ is necessary to get no contradiction with experiment.

There will often be the case that $U^{\alpha i}$ is very large in certain regions. In these cases we have introduced in Sect. 4.3.3 the fundamental domain $G$ as a part of the application domain $A_p$. If $\Sigma_\beta$ is a better idealization than $\Sigma_\alpha$ in the described sense, i.e., if $U^{\beta i}$ is much smaller than $U^{\alpha i}$, then we get, from the definition of $G$, that $G_\alpha$ must be smaller than $G_\beta$. Sometimes it can be that $G_\beta \equiv A_{p_\beta}$, but $G_\alpha$ is only a part of $A_{p_\alpha} \equiv A_{p_\beta}$, and $G_\alpha$ is already determined by $\widehat{U}_\nu$.

The formulation of the relation between two physical theories given above may not seem sufficiently precise. But this is the mathematically exact formulation of the intuitive formulation that $\Sigma_\beta$ resembles the structure of reality more than $\Sigma_\alpha$. That this mathematical formulation needs the inaccuracy sets, which cannot be laid down, is a consequence of our physical theories, where we use the idealizations $\Sigma_\alpha, \Sigma_\beta$, which are only inaccurate pictures of reality. Some physicists try to describe this situation in the following way: $\Sigma_\alpha$ is a hypothesis about the structure of reality; we take this $\Sigma_\alpha$ as an exact picture, and if we state later that this is not the case, we try to formulate a new hypothesis $\Sigma_\beta$ and take this again as being exact. But we do not know what we must do if a $\Sigma_\alpha$ or $\Sigma_\beta$ is exact, even if we know in advance that the hypotheses are all false or, even better, only inaccurate pictures. We prefer the formulation of the relation between $\Sigma_\alpha$ and $\Sigma_\beta$, which we have given above. Since this relation is the essential starting point for all other relations between physical theories, we will now give some simple examples.

As a first example we take a theory for the equilibrium states of a gas. The application domain $A_p$ consists of a mole of a particular gas and the values of pressure, volume, and temperature in equilibrium. Thus we have for $\Theta$ the set $\overline{\mathbb{R}} \times \overline{\mathbb{R}} \times \overline{\mathbb{R}}$ with a relation $\bar{s} \subset \overline{\mathbb{R}} \times \overline{\mathbb{R}} \times \overline{\mathbb{R}}$ where we take the first $\overline{\mathbb{R}}$ as the values $p$ of pressure, the second $\overline{\mathbb{R}}$ as the values $V$ of volume, and the third $\overline{\mathbb{R}}$ as the values $T$ of temperature. We take as idealizations $MT_\Sigma$:

$MT_{\Sigma_\alpha} : s_\alpha \subset \mathbb{R} \times \mathbb{R} \times \mathbb{R}$ represents the ideal gas state
equation $pV = RT$;

and

$MT_{\Sigma_\beta} : s_\beta \subset \mathbb{R} \times \mathbb{R} \times \mathbb{R}$ represents the van der Waals state
equation $(p + a/v^2)(V - b) = RT$.

Without experiments we cannot know whether one of these two equations represents "better" than the other the experimental results. But from experiments we know that we must take for $\mathbb{R} \times \mathbb{R} \times \mathbb{R}$ in $MT_{\Sigma_\alpha}$ much greater inaccuracy sets to get $\Delta_\alpha$ than in $MT_{\Sigma_\beta}$ to get $\Delta_\beta$, so that $MT_{\Delta_\alpha}$ and $MT_{\Delta_\beta}$ give no contradiction with experiment. Especially for $\Delta_\alpha$ we have to choose in $\mathbb{R} \times \mathbb{R} \times \mathbb{R}$, for small $V$ and small $T$, very large inaccuracy sets. Thus we get that the fundamental domain $G_\alpha$ of the theory $PT_\alpha$ is a region where $V$ and $T$ are not too small, i.e., a not too dense gas for higher temperatures.

Our example of the equilibrium values of $p, V, T$ for a gas is nothing other than a three-dimensional case of two-dimensional cases well known in experimental physics:

We consider a relation $\bar{s} \subset \overline{\mathbb{R}} \times \overline{\mathbb{R}}$ and, as an idealization, a relation $s \subset \mathbb{R} \times \mathbb{R}$ which represents a real function $y = f(x)$ (see Fig. 5.1).

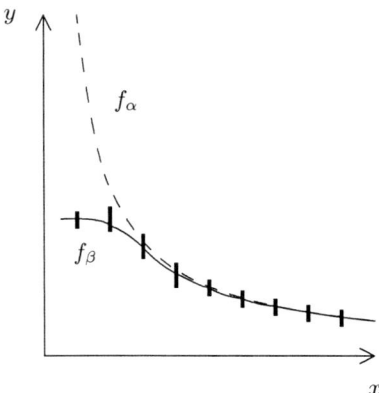

**Fig. 5.1.** Ideal gas ($f_\alpha$) and van der Waals ($f_\beta$) curves

We may have two theories, where the first gives the function $y = f_\alpha(x)$ and the second $y = f_\beta(x)$ (see Fig. 5.1). The little black rectangles represent the measured values with their intervals $J$ (the rectangles) of "errors of measurement." We see that the function $f_\beta$ is a "better" idealization of the measured values than the function $f_\alpha$. How do we describe this "better" according to our method, i.e., in a mathematically formulated way?

Figure 5.1 shows that the errors of measurement are so large that we cannot see the deviations of the idealization $f_\beta$ from the reality, i.e., we can choose very small $U^{\beta i}$. Thus $U^{\alpha f} = \widehat{U} \circ U^{\beta i} \approx \widehat{U}$ is the deviation of $f_\alpha$ from $f_\beta$, so that the environment of $(f_\alpha)_{\widehat{U}}$ includes $f_\beta$. $\widehat{U}$ describes how much $f_\alpha$ deviates from $f_\beta$ and (in the approximation of $U^{\beta i}$) how much $f_\alpha$ deviates from the reality. This deviation of $f_\alpha$ is much larger than the small deviation of $f_\beta$ from the reality, that has not yet been observed by the state of measurement errors until now. It may be that by future developments we will discover that $f_\beta$ also deviates from the reality but much less than $f_\alpha$. In the example, we describe that the idealization of the surface of the earth by an Euclidean geometry $\Sigma_\alpha$ is not as good as the idealization $\Sigma_\beta$ by the geometry of the surface of the globe.

## 5.2 Relations Between Two *PT*s with a Common Part of an Application Domain

What do we mean by the assertion "the application domain $A_{p_\alpha}$ of the theory $PT_\alpha$ is a part of the application domain $A_{p_\beta}$ of the theory $PT_\beta$"?

The equivalence $A_{p_\alpha} \equiv A_{p_\beta}$ was characterized by the following: every $\widetilde{A}$ of $PT_\alpha$ is also an $\widetilde{A}$ of $PT_\beta$ and vice versa. Now we define ⌈ $A_{p_\alpha}$ is a part of $A_{p_\beta}$ ⌉, briefly $A_{p_\alpha} \subset A_{p_\beta}$, if every $\widetilde{A}$ of $MT_\alpha$ is also an $\widetilde{A}$ of $MT_\beta$.

This is only the case if $B_{l_\beta}$ uses all concepts of $B_{l_\alpha}$, i.e., if $B_{l_\beta}$ uses all propositions, stating properties of objects and relations between objects, of $B_{l_\alpha}$. This does not mean that the basic properties of $B_{l_\alpha}$ are also basic properties of $B_{l_\beta}$. The properties and relations of $B_{l_\alpha}$ can also be logical combinations by "and" and "not" of those of $B_{l_\beta}$. Concerning the basic properties, we postulate in addition that an object which has a basic property of $B_{l_\alpha}$ also has a basic property of $B_{l_\beta}$. How can we formulate all this in the language of $\Theta_\alpha$ and $\Theta_\beta$?

We postulate as a condition for $A_{p_\alpha} \subset A_{p_\beta}$: To every property $p_\alpha(y)$ of $\Theta_\alpha$ corresponds a property $p_\beta$ of $MT_{\Theta_\beta}$, to every relation $r_\alpha(x, y, \ldots)$ of $\Theta_\alpha$ corresponds a relation $r_\beta$ of $MT_{\Theta_\beta}$.

In the framework of $\Delta_\alpha$ and $\Delta_\beta$ we can formulate this in the following way: To every base set $\overline{M}_\alpha$ of $\Delta_\alpha$ corresponds a set $E_\beta$ of $MT_{\Delta_\beta}$, to every structure term $\overline{s}_\alpha \subset T(\overline{M}_1, \ldots)$ of $\Delta_\alpha$ corresponds a term $U_\beta \subset T(E_1, \ldots)$ of $MT_{\Delta_\beta}$.

This correspondence is defined by the basic languages $B_{l_\alpha}$ and $B_{l_\beta}$: $E_\beta$ describes in $MT_{\Delta_\beta}$ the basic properties of $B_{l_\alpha}$, $U_\beta$ describes in $MT_{\Delta_\beta}$ the properties and relations of $B_{l_\alpha}$.

In Sect. 4.2 we described what was meant by "the terms $U_\beta$ over the base $E_\beta$ form a structure of species $\Delta_\alpha$ in $MT_{\Delta_\beta}$." We now define

If "the terms $U_\beta$ over the base $E_\beta$ form a structure of species $\Delta_\alpha$ in $MT_{\Delta_\beta}$," then we say that ⌈ $MT_\beta$ is "richer" than $MT_\alpha$ ⌉.

One sees immediately that, in the special case where

$$E_\beta = \overline{M}_\beta \quad \text{and} \quad U_\beta = \overline{s}_\beta \text{ (of } \Delta_\beta\text{)},$$

we obtain the case $A_{p_\alpha} \equiv A_{p_\beta}$ dealt with in Sect. 5.1. Thus the above definition of a richer $PT$ is a generalization of the definition in Sect. 5.1. But it is not necessary to repeat all the considerations of Sect. 5.1, since they are essentially the same, but with $\Delta_\alpha$ substituted by $(E_\beta, U_\beta)$. Whether "the terms $U_\beta$ over the base $E_\beta$ form a structure of species $\Delta_\alpha$ in $MT_{\Delta_\beta}$" essentially depends on the inaccuracy sets (as described in Sect. 5.1).

Instead of repeating the considerations of Sect. 5.1 we will only give some examples of physical theories:

- $MT_\alpha$ describes hydrogen atoms by quantum mechanics, $MT_\beta$ describes all atoms (in this case we have to know and to use in $MT_\beta$ the Pauli's exclusion principle, but not in $MT_\alpha$);
- $MT_\alpha$ describes atoms, $MT_\beta$ describes atoms and molecules;
- $MT_\alpha$ is a thermodynamic of gases only, $MT_\beta$ is a thermodynamic of gases with different phases: gases, fluids, and rigid bodies.

In the development of new theories, one often has a chain of theories ⌈ $MT_{\alpha_2}$ is richer than $MT_{\alpha_1}$ ⌉, ⌈ $MT_{\alpha_3}$ is richer than $MT_{\alpha_2}$ ⌉, ... until one has reached a general theory $MT_\beta$. One then often "forgets" the various steps of $MT_\alpha$ in order to reach $MT_\beta$, because one can introduce in $MT_\beta$ the various $(E, U)$ instead of $\Delta_\alpha$. This is a particular case of the general method to introduce "approximation theories." Before we generalize again the concept of a "richer" theory, it is necessary to consider another relation between physical theories: ⌈ $MT_\gamma$ is a "pre-theory" of $MT$ ⌉.

## 5.3 Pre-theories

Why do we use pre-theories and not just "one" theory for all physical problems? There are two reasons. The first is that we do not have a single theory for all of physics. Also, the authors of this book believe in the "existence" of this "single theory," but only as a *goal* that we have not yet reached and perhaps never will. The second and most important reason is the intention to formulate theories in such a way that is not too difficult to work with these theories. As, e.g., Kepler, who was interested in the motion of the planets, used as a "pre-work" the measurements of Tycho Brahe as a result of a space–time theory. He used the space–time positions as determined by a "pre-theory." Indeed, practically all physical theories are formulated in such a way, i.e., pre-theories are used.

The concepts of the basic language $B_l$ of a $PT$ are given either immediately, by the description of what we observe by sensory perception (i.e., using our senses), or are defined with the help of pre-theories. Terms of a pre-theory (not necessarily base terms!) are taken in $B_l$ as designating "known" concepts used to formulate the properties and relations of the $B_l$ of $PT$.

This use of pre-theories to get some concepts of the basic language contains two problems:

1. *A general problem.* For a $PT \equiv A_p \leftrightarrow B_l(\text{cor})MT_\Delta$ we can take terms other than only the base and structure terms of $\Delta$ to define new concepts used in an extension of $B_l$. This problem is described in Sect. 6.3 where we ask how we can obtain by $MT_\Delta$ new knowledge about reality, going beyond the knowledge of the pre-theories used in $B_l$.
2. *A particular problem.* If $PT_\gamma \equiv A_{p_\gamma} \leftrightarrow B_{l_\gamma}(\text{cor})MT_{\Delta_\gamma}$ is a pre-theory of $PT$, we have to select in $MT_{\Delta_\gamma}$ the terms which we will use in $PT$ and define terms by which we designate the related concepts in $B_l$.

The definition of words is essential for the language of physics and therefore for work with physical theories. One can in this way avoid repeating the history of the experimental measurements of these "new" concepts. For the use of a $PT$ we can forget "how" we can measure (respectively what we have measured) the properties and relations of $B_l$, i.e., how we can get $\widetilde{A}$ for this $PT$.

Let us describe more precisely the connection between the terms designating concepts in a pre-theory $PT_\gamma$ and the terms designating concepts in the theory $PT$. Let '$h_\gamma$' be a term, designating a concept "$h_\gamma$" of $MT_{\Delta_\gamma}$, selected according to point (2) above. Since we can represent properties and relations by sets (see Sect. 3.2.2), we will assume that "$h_\gamma$" is a set.

Let '$c$' be the new term designating the new concept "$c$" of $B_l$ corresponding to "$h_\gamma$." Let '$k$' be the new term designating the new concept "$k$" of $\Theta$ corresponding to "$c$." We then get a correspondence "$h_\gamma$" $\leftrightarrow$ "$c$" $\leftrightarrow$ "$k$" by which we can change the form of the $PT$.

We go from a structure $\Delta$ to a new extended structure $\Delta_{\text{ex}}$ by adding the base terms, the structure terms, the axioms of $\Delta_\gamma$, *and* by identifying "$k$" with "$h_\gamma$." As basic language $B_{l_{\text{ex}}}$ of this theory we have to use the basic language $B_{l_\gamma}$ plus that part of $B_l$ which does *not* use the concepts defined by $PT_\gamma$ (i.e., not the concept "$c$" defined above by "$h_\gamma$" $\leftrightarrow$ "$c$" $\leftrightarrow$ "$k$").

This new theory $PT_{\text{ex}}$ with $MT_{\Delta_{\text{ex}}}$ shows how the pre-theory $PT_\gamma$ is working in $PT$. But this advantage is outweighed by the disadvantage that in most cases $MT_{\Delta_{\text{ex}}}$ will be much more complicated than $MT_\Delta$. Therefore one will not use $MT_{\Delta_{\text{ex}}}$ for the physical work, but $MT_\Delta$ or perhaps even, instead of $MT_\Delta$, a simpler approximation theory $MT_{\Delta_{\text{appr}}}$ (see Sect. 5.5).

Since $PT_{ex}$ *only* describes the details of how $PT_\gamma$ is working in $PT$, $PT_{ex}$ describes all of the real structures which describe $PT$, but namely only a little more the working of $PT_\gamma$ in $PT$. We say therefore that

⌈ $PT_{ex}$ is "richer in content" than $PT$ ⌉

and write

$PT_{ex} \succ PT$.

Such extended $MT_{\Delta_{ex}}$ are only of interest for the questions dealt with in Sect. 5.4, and in the context concerning the question of how the totality of all physical theories is founded on a reality, the description of which is possible without any physical theory. Until now this problem has not been solved. Therefore we will give a description of this problem, which we are convinced can be solved.

There exists the belief that physics cannot be founded on the basis of immediately recognizable realities as, e.g., a stone onto which one has stumbled. Some people believe that the first impression of a hard and rigid stone was refuted by physics; that physics detected that the stone is essentially only an empty space since neutrinos pass this stone with very few exceptions. But this is a misunderstanding. The stone remains as hard and as rigid as before; what has changed with physics is our knowledge of reality. Our vision of physics is that we have to begin with immediately recognizable facts and, on this basis, learn by physical theories more and more about the reality. We never have to refute what we have already detected, except when we have made a mistake in the application of the method of physical theories described here in this book.

This method is based on immediately recognizable facts, which can be denoted by sentences using concepts of a basic language, concepts which we learn by demonstration and not by physical theories. This *initial basic language* is denoted by $B_{l_i}$. The first theories have as a basic language $B_{l_i}$ or a part of $B_{l_i}$. Then one can use these theories as pre-theories for future theories, and some of these theories can again be used as pre-theories for other theories.

Physics is therefore defined by the choice of $B_{l_i}$ and by the method $PT \equiv A_p \leftrightarrow B_l(\text{cor})MT_\Delta$ of a physical theory $PT$.

The fact that we use as $B_{l_i}$ only a part of those facts that we can describe without any physical theory is essential for that part of science which we call physics. The method $A_p \leftrightarrow B_l(\text{cor})MT_\Delta$ can also be applied to the nonphysical part of reality. We have shown this here by Example C (see Part II), and one can read such applications also in [11]. But what do we choose as $B_{l_i}$?

For instance, the sentence 'this tree has green leaves' is certainly a sentence of the initial language, since the concepts "tree," "has," "green," and "leaf" have been learnt without any physical theory. Concepts learnt by physical theories are, e.g., "electron," "H-atom," "molecule," and "electromagnetic field." What is a tree, a leaf, and green "have been learnt from demonstration." Nevertheless, we do not use the above sentence for the $B_{l_i}$ of physics. So what do we allow as sentences of $B_{l_i}$?

The choice of concepts has indeed changed during the development of physics. Even more, one did not formulate the problem of $B_{l_i}$ or clear up the concepts used in $B_{l_i}$, so this has often produced misunderstandings. A famous misunderstanding was the use of the concept "color" (such as green, red, and yellow) by physicists (such as Newton) or by artists (such as Goethe, who also painted pictures). Newton and Goethe have used the same word 'color' to designate different concepts. Similarly, there was the misunderstanding in the use of the word 'motion' by Galilei, whose meaning did not have the same sense as that of the philosophers in the church who used this word to designate a concept of "change," which must have a "cause." Today we have no difficulty in using the sentence 'the sun has risen' with the sentence 'the earth is rotating'.

Aware of these complications, one has during the development of physics reduced more and more the basic language $B_{l_i}$. This step-by-step reduction is illustrated by various words, describing various parts of physics: the 'optic' was that part which dealt with 'light', the 'acoustic' was that part which dealt with 'sound', the 'thermodynamic' was that part where 'warm' and 'cold' play an essential role. But all these particular parts disappeared as "particular" parts, i.e., as parts with particular parts of $B_{l_i}$. These parts were absorbed into other parts, where in $B_{l_i}$ such words as 'light', 'sound', 'warm', 'green' and 'red', etc. do not appear. One had reduced $B_{l_i}$ and one continues to reduce it today. We move more and more toward digital events; instead of looking at scales, one today looks at "numbers."

We might also mention other concepts which we do not accept as concepts of $B_{l_i}$. They are the concepts "now," "past" and "future." These concepts are essential in our daily life, but they do *not* appear in any physical theory!

But there is no doubt that until now we have no exact definition of $B_{l_i}$. Nevertheless, it seems to cause no essential difficulties between the physicists about what is ascribed to physics. We have to consider also chemistry as being a part of physics, since we have no other concepts used in the basic language $B_l$ of chemistry other than those of physics, and we have the same method of theories, even if in chemistry the mathematical structure of these theories is not as elaborated as it is in physics.

The range of physics is determined by $B_{l_i}$ and the method $PT \equiv A_p \leftrightarrow B_l(\text{cor})MT_\Delta$ of theories. Since today we have reduced the $B_{l_i}$, we have also reduced the region $W$ of realities which we can describe by physics. This reduction does not mean that there are no other realities other than those of $W$. On the contrary, we will come back to this problem in Chap. 6 where we will discuss how we can obtain more and more knowledge about $W$.

## 5.4 Relations Between *PT*s with Different Application Domains

One has for instance Newton's gravitational theory which uses Newton's space–time theory as a pre-theory. One feels intuitively that Einstein's gravitational theory is better, i.e., richer, than Newton's theory. But Einstein's theory changes also the pre-theory of Newton's theory: the space–time theory with an Euclidean space and an absolute time. Can we give to this intuitive meaning a more exact description?

We may have two theories $PT_\alpha$ and $PT_\beta$ where we have the intuitive meaning that ⌈ $PT_\beta$ is richer than $PT_\alpha$ ⌉. But the pre-theories of $PT_\alpha$ are not also pre-theories of $PT_\beta$, and so the definition of "richer in content" given in Sects. 5.1 and 5.2 cannot be applied.

There may be one pre-theory $PT_\gamma$ of $MT_\alpha$ which is not a pre-theory of $MT_\beta$. But there may be a pre-theory $PT_\delta$ of $MT_\beta$ which is richer than $PT_\gamma$ in the sense of Sect. 5.1 or Sect. 5.2, or $PT_\beta$ itself is richer than $PT_\gamma$, i.e., $PT_\delta \succ PT_\gamma$ or $PT_\beta \succ PT_\gamma$.

We will discuss at first the case where $PT_\beta \succ PT_\gamma$. Then we can construct the theory $PT_{\alpha\mathrm{ex}}$ (as described in Sect. 5.2) by the construction of $MT_{\Delta_{\gamma\mathrm{ex}}}$. We then have $PT_{\alpha\mathrm{ex}} \succ PT_\alpha$ (in the sense of Sect. 5.2). (If there are more than one pre-theory $PT_\gamma$ of $PT_\alpha$, which is not a pre-theory of $PT_\beta$, then one must do the same thing with these theories as shown above with $PT_\gamma$ in order to obtain $PT_{\alpha\mathrm{ex}}$.) Then the theory $PT_{\alpha\mathrm{ex}}$ only contains pre-theories which are also pre-theories of $PT_\beta$. Thus we can ask whether $PT_\beta \succ PT_{\alpha\mathrm{ex}}$. We therefore define

⌈ $PT_\beta$ is richer than $PT_\alpha$ ⌉

and write

$PT_\beta \succ PT_\alpha$.

This relation is valid for $PT_\alpha$ as Newton's gravitational theory, and $PT_\beta$ as Einstein's gravitational theory. In this case, $PT_\gamma$ is the pre-theory of Newton's space–time theory, and the space–time theory of Einstein is contained in $PT_\beta$. We have to formulate the pre-theory $PT_\gamma$ in such a way that we get $PT_\beta \succ PT_\gamma$. We do this by the formulation of $B_{l_\gamma}$ in such a way that $B_{l_\gamma}$ is a part of $B_{l_\beta}$. As $B_{l_\gamma}$ we introduce the description of *local* space–time reference systems, which are initial systems, i.e., systems without forces which depend on masses. The global connection of these local systems is described in $PT_\beta$ *and* $PT_\gamma$ by an affine connection. The difference is that this affine connection is determined in $PT_\gamma$ by the masses (as described in [14]), and in $PT_\beta$ by a four-dimensional metric that fulfills Einstein's equations. The common space–time pre-theory of $PT_\gamma$ and $PT_\beta$ are the local initial systems and their affine connections; the difference between the two are the axioms for these connections in $PT_\gamma$ and $PT_\beta$. To show that $\lceil$ $PT_\beta$ is richer than $PT_\gamma$ $\rceil$, i.e., to show that $PT_\beta$ describes the reality more precisely than $PT_\alpha$, can only be demonstrated by experiments. Why many physicists (such as Einstein himself) believed without knowing the results of experiments that $PT_\beta \succ PT_\alpha$, whilst others have principally refuted Einstein's theory $PT_\beta$ is very amusing, and shows that arguments other than physical ones can play a decisive role in deciding subjectively whether one acknowledges a theory or not. Arguments that acknowledge Einstein's theory without new experiments are discussed in Sect. 5.5.

One can generalize the above definition of $PT_\beta \succ PT_\alpha$ to the case where not $PT_\beta \succ PT_\gamma$, but where only $PT_\delta \succ PT_\gamma$ (see the beginning of Sect. 5.4). Then one can construct with $PT_\delta$ the theory $PT_{\beta\mathrm{ex}}$. Then $PT_{\beta\mathrm{ex}} \succ PT_\gamma$. If $\lceil$ $PT_{\beta\mathrm{ex}}$ is richer than $PT_{\alpha\mathrm{ex}}$ $\rceil$, we say that $\lceil$ $PT_\beta$ is "essentially richer" than $PT_\alpha$ $\rceil$.

## 5.5 Approximation Theories

One could perhaps think that most of the efforts of theoretical physicists consist in finding richer and richer theories. But this is not the case. Most of the efforts consist in finding "poorer" theories, or so-called approximation theories. But why do we seek such theories?

Not because they are less rich than a well-known and very good theory! We seek such approximation theories for mathematical reasons. A very good theory is often mathematically so complicated that it seems hopeless for solving particular problems as, e.g., the structure, the binding energy, and the spectral lines of a molecule.

If we have a good theory $PT \equiv A_p \leftrightarrow B_l(\mathrm{cor})MT_\Delta$, we seek to find another approximation theory $PT_\mathrm{appr} \equiv A_{p_r} \leftrightarrow B_{l_r}(\mathrm{cor})MT_{\Delta_\mathrm{appr}}$ which is much simpler for applications in an eventually also reduced application domain $A_{p_r} \subset A_p$, or better in a very reduced fundamental domain $G_r \subset A_{p_r}$ and $G_r \subset G$ (with $G$ as the fundamental domain of $PT$). We seek to find a much simpler approximation theory $PT_\mathrm{appr}$ which is less rich than $PT$: $PT_\mathrm{appr} \succ PT$, a relation which we have described in Sect. 5.1 to Sect. 5.4. It is not possible to provide a method for finding a physical theory, neither for a richer nor for a simpler approximation theory. All the imaginative and intuitive ideas can be used to create new theories. Therefore we could finish by describing the relation $PT_\mathrm{appr} \succ PT$ since this relation is already described in Sect. 5.1 to Sect. 5.4.

Let $PT_\alpha \equiv A_{p_\alpha} \leftrightarrow B_{l_\alpha}(\mathrm{cor})MT_{\Delta_\alpha}$ be a given theory where $\Delta_\alpha$ is known, i.e., that also the inaccuracy sets for $\Delta_\alpha$ are known. We seek relative to the given theory $PT_\alpha$ a richer theory $PT_\beta \equiv A_{p_\beta} \leftrightarrow B_{l_\beta}(\mathrm{cor})MT_{\Delta_\beta}$. The inaccuracy sets for $\Delta_\beta$ are to be found by experiment and cannot be deduced from $\Delta_\alpha$. If we seek, relative to a given theory $PT \equiv A_p \leftrightarrow B_l(\mathrm{cor})MT_\Delta$, a simpler approximation $PT_\mathrm{appr} \equiv A_{p_r} \leftrightarrow B_{l_r}(\mathrm{cor})MT_{\Delta_\mathrm{appr}}$, then the inaccuracy sets for $\Delta$ must be very small. Then we have to find the inaccuracy sets for $\Delta_{\text{appr}}$. They must be at least of such a size that ⌈ $\Delta$ is richer than $\Delta_{\text{appr}}$ ⌉, i.e., that $\Delta$ is a structure of the species $\Delta_{\text{appr}}$ in $MT_\Delta$. In this case, the inaccuracy sets for $\Delta_{\text{appr}}$ could be found mathematically. But the physicists seldom do this, much to the disappointment of the mathematicians. The principal aim of physicists is to work with approximation theories in order to obtain good applications (in the new techniques). Therefore, there is competition (also regarding money) for the development of this work: to find richer and richer theories, and to find richer and richer applications (especially in the domain of high technology) by the construction of more usable, but not quite so rich, approximation theories. Amongst the multitude of such approximation theories, we mention only the approximation theories by which we try to get an insight into the structure of molecules, and also into their behavior when collisions occur.

Often we use theories that are not so rich, such as approximation theories, with the aim of later finding a richer and also more complex theory. For instance, if we imagine that Einstein's gravitational theory $PT_\beta$ would have been found before Newton's gravitational theory, there would have been an urgent need to develop Newton's theory as an approximation theory of Einstein's theory. Thus we can also read Sect. 5.4 as describing the relation between an approximation theory and the richer original theory.

But we shall not forget that physics is not only a means for making better technical developments. If a good technical development is useful for the life of human beings today, we should also be interested in the structure of the reality in which we live. Therefore we are also interested in the structure of the totality of physical theories, a problem which we will try to discuss in the following section.

## 5.6 The Network of $PT$s

In the totality of physical theories, there exists the essential relation ⌈ $PT_1$ is richer than $PT_2$ ⌉, denoted by $PT_1 \succ PT_2$. This relation has a similarity to the mathematical order relation $>$. But the totality of physical theories is not a set in the mathematical sense. Nevertheless, we can discuss the relation $\succ$ similar to $>$ if we do not use such mathematical relations as $\forall$ and $\exists$. We have to avoid speaking of "all" theories, and we must use expressions such as "there exists a theory" only if this theory exists today.

The relation $\succ$ is similar to $>$, since we have seen previously in Sects. 5.1 to 5.4 that from $PT_1 \succ PT_2$ and $PT_2 \succ PT_3$ follows $PT_1 \succ PT_3$. We call the totality of physical theories with this relation $\succ$ the "network of physical theories." This network is changed if we find new "richer" theories and construct new "not so rich" approximation theories. The structure of this network is very interesting but not very well known. Nevertheless, we will try to discuss some of the features of this network.

## 5.6 The Network of $PT$s

There is at first the question of whether this network is connected, i.e., whether one can go from one theory $PT_1$ to another theory $PT_2$ step by step, with relations $\succ$, or whether there are different separate parts within this network?

One could perhaps think that the network is connected, because we have for all theories the same initial basic language $B_{l_i}$. But this argument is not stringent. The language $B_{l_i}$ itself may be "not connected" in the following way: We say that two properties in $B_{l_i}$ are connected if it is possible "to go" from one property to another, step by step, by relations of $B_{l_i}$. If in $B_{l_i}$ there are two separate parts, we can also have two different pre-theories which are not connected, i.e., where $\Theta_1$ has no common part (i.e., common interpreted part) with $\Theta_2$.

We can join together two theories $PT_1$ and $PT_2$ as one $PT$, by taking together the structures $\Theta \equiv \langle \Theta_1, \Theta_2 \rangle$, and by taking together the axioms $\Delta \equiv \langle \Delta_1, \Delta_2 \rangle$; we postulate for every $PT$ that this $PT$ is connected in the following way.

The structure $\Theta$ of $PT$ cannot be separated into two parts $\Theta_1$ and $\Theta_2$, so that there are no relations $\bar{s}$ of $\Theta$ with a typification $\bar{s} \subset \overline{M}_1 \times \overline{M}_2 \times \cdots$ such that $\overline{M}_i$ of $\Theta_1$ *and* $\overline{M}_k$ of $\Theta_2$, and that there are only axiomatic relations of $\Delta$ which contain only relations between $\overline{M}_i$ and $\bar{s}$ of the same $\Theta_1$ or the same $\Theta_2$.

The language $B_{l_i}$ does not prevent the fact that there are physical theories which are not connected.

*One can only decide whether the network of physics today is connected or not by the explicit construction of this network. But this has not been done until now. Therefore, we can only fix our supposition: We suppose that physics today (including chemistry) is a connected network. For this structure, the restriction of the language $B_{l_i}$ is essential.*

But much more interestingly than the connection of the network of physical theories is the question of whether the network is (hypothetically) directed.

We say that the network is hypothetically directed if, for two theories $PT_1$ and $PT_2$, a third theory $PT_3$ can be developed (or founded) with $PT_3 \succ PT_1$ and $PT_3 \succ PT_2$.

The supposition that the network of physical theories is hypothetically directed can be outlined by the following explanation.

It is possible to develop a theory $PT$ for which $PT \succ PT_\nu$ for all theories $PT_\nu$. The richest theory $PT$ is often called a "world formula." Some people think that they have already discovered this "world formula." We do not believe that this "richest" theory exists. Since we believe that physics is principally "finite" (see Sect. 3.2.4), the richest theory $PT$ should be a theory without idealization, e.g., a purely finite mathematical theory. Perhaps we will never find this "best" theory $PT$, but this best theory remains an objective which we try to approach.

Steps in this approach procedure are very fruitful. Therefore we will try to analyze such steps.

We often have the situation given by the diagram of Fig. 5.2.

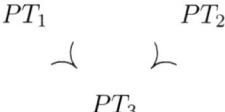

**Fig. 5.2.** Relations between three $PT$s

Here $PT_3$ can be, e.g., an approximation theory for $PT_1$ and $PT_2$.

Such a situation is, e.g., given by the following: $PT_3$ is Newton's mechanics and space–time theory *without* gravitational forces. $PT_1$ is $PT_3$ plus Newton's gravitational law. $PT_2$ is the special relativity theory and mechanics without gravitation. This was the situation after the discovery of Einstein's special relativity theory. Many people tried to unify the two theories $PT_1$ and $PT_2$ by finding a form of gravitational law that is invariant under the special relativity theory. But the right solution was $PT_4$ as Einstein's gravitational theory. Einstein's solution for $PT_4$ was for many physicists (and also for the authors of this book) so natural that they believed in this theory without any knowledge of new experiments, which showed later that indeed $PT_4 \succ PT_2$. If the world is not stupidly constructed, $PT_4$ has to be the "better" theory, i.e., the more realistic theory. With $PT_4$ as Einstein's theory we have the diagram of Fig. 5.3.

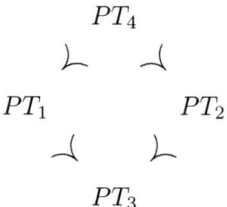

**Fig. 5.3.** Relations between various space-time theories

The diagram of Fig. 5.2 should be a means for seeking a theory $PT_4$ in order to obtain the diagram of Fig. 5.3. But it can also be that the history of physics shows us at first the theory $PT_4$, and that we try afterwards to obtain the theories $PT_1, PT_2$, and $PT_3$ as approximation theories of $PT_4$. If, e.g., at first we had Einstein's gravitational theory, it would be necessary to develop Newton's theory as an approximation, which is good enough to calculate the motion of the planets, of the moon, and of our spaceships.

# 6
# Real and Possible as Physical Concepts

Since the beginning of our description of the structure of a physical theory in Sect. 3.2, we started with the application domain $A_p$ as a part of "reality," as that part which we can describe by the basic language $B_l$. Thus we presumed that we could gain for a $PT$, by immediate observations and by pre-theories, a text $\widetilde{A}$ as a description of a part of $A_p$; only a part, since for practically all $A_p$ we have never established "all" the realities of $A_p$. In Sect. 5.3 we have seen that with an analysis of the pre-theories we can go back from the language $B_l$ to a language $B_{l_i}$ which does not use any pre-theory. Such application domains $A_{p_i}$, with languages $B_{l_i}$, are the starting points of all of physics. To a part of $A_{p_i}$, formulated in a language $B_{l_i}$, corresponds a text $\widetilde{A}_i$ which we "presume" to be a description of a part of the "reality." We "presume" such immediately observed realities and do not base them on philosophical considerations. We presume, e.g., that there is a "real" cup of tea if we have observed this cup of tea on a table.

As already mentioned in Sect. 5.3, we do not use in $B_{l_i}$ all of the concepts that we use in everyday language. So we have "excluded from physics" the concepts of color and sound, but also the concepts of "past," "present," and "future." Thus, these three concepts are not used in physical theories! We have in some physical theories only "time scales" as given by clocks, *without* introducing a concept specifying which of these "time scale values" is "now" (the "present" value).

This remark is very important if we compare the physical concept of "real" with a philosophical one. There are philosophical concepts of "real" for which the time plays an essential role! In physics we consider something as real independently of time. It doesn't matter whether it "was" or it "is" or it "will be"! The main point of a physical theory is *not* to have an instrument for predicting what "will be." Nevertheless, a physical theory can sometimes be used to make predictions, and also to state realities at a time where we have not made observations!

There is in no $PT \equiv A_p \leftrightarrow B_l(\text{cor})MT_\Delta$ a concept of what we call "now" in our consciousness. In the text $\widetilde{A}$ we can only write what we *know* until now. In our concept of a physical theory, our "knowledge" is *not* a part of $MT_\Delta$; it intervenes only in testing that $MT_\Delta \overline{A}$ is without contradiction. Our knowledge appears only in the *additional* text $\overline{A}$.

There are some who believe that the mathematical theory $MT$ of quantum mechanics describes changes of our knowledge, i.e., that one has to use concepts such as "I know ..." for the interpretation of the mathematical part $MT$ of quantum mechanics. We do not believe this, i.e., we are convinced that we can also introduce in quantum mechanics an objective concept of "real" which does not depend on our knowledge: $\widetilde{A}$ and $\overline{A}$ only describe the knowledge of an objective reality of a part of $A_p$, and $MT$ does not depend on the extent of our knowledge $\overline{A}$. See the description of an axiomatic basis $MT$ for quantum mechanics in [8].

There are many other physicists who believe in the reality of fairy tales if the used $\Sigma$ (the fairy tales) generates a theory $PT$ which is not in contradiction to experiment. But this "to not be in contradiction with experiment" is certainly not enough to "prove" the reality, i.e., to find a conviction of the reality of a fairy tale. It is, e.g., not difficult to add to a well-tested theory some fairy tales without influencing the physical content of the theory. We can, e.g., add to Einstein's gravitational theory a fairy tale of a Newton's space–time structure, i.e., one Euclidean three-dimensional space and *one* time scale; then one can take Einstein's metric field $g_{\mu\nu}$ as a field in this space and time. One can "believe" that the Newton's space–time structure describes a "real" structure (but one hidden to physical measurements) in which the Einstein's field $g_{\mu\nu}$ describes the behavior of clocks and the motion of bodies (i.e., the "forces"). A belief in such a "real" "absolute" space–time structure of Newton's form cannot be based on philosophy.

There are other historical examples of fairy tales as, e.g., the "ether" as a body, the elastic motion of which are the light waves. This fairy tale was abandoned not because of contradictions with measurements of light, but because the new generation of physicists did not "believe" in the reality of this ether, and therefore had no desire to work on this fairy tale, which would have described all experiments about light (a task which is not impossible).

Until today, quantum mechanics has been a rich source (an "Eldorado"?) for the invention of fairy tales. A well-known fairy tale is the famous search for "hidden variables." The word "hidden" means here exactly that these variables describe no physical reality, i.e., that these variables are only a fairy tale. But why do some physicists seek such fairy tales? Why are they not satisfied with a theory formed according to the methods given in this book?

Physicists are not satisfied because they believe in a certain "a priori" structure of reality, what quantum mechanics (as developed according to the methods of this book, i.e., as an axiomatic base described in [8]) cannot provide.

But there are also other fairy tales in quantum mechanics as, e.g., the following: One begins with two concepts not defined physically but only mathematically: the concept of a "state" defined as a vector of a Hilbert space and the concept of a "superposition principle" defined by the addition of vectors in a Hilbert space. This is a typical example of a fairy tale since one covers the mathematical terms of $MT_\Sigma$ with words (e.g., a state) which have in other physical theories, such as thermodynamics or Newton's point mechanics, a *well-defined* significance. One tries in this way to give a "feeling" of a significance: the words shall have a "similar" significance, without saying in what this similarity consists. A lot of effort has been made to develop on this basis a "quantum mechanics," and not to let oneself be deferred by such incredible features as the so-called "collapse of wave packets" at the "moment" of a measurement.

We do not believe in this fairy tale and especially not in the so-called "collapse of wave packets." Why should we believe in such a thing as being real? A reason could be the belief in some "a priori" structures of the physical reality. Another reason could be the desire to make quantum mechanics a description of an "isolated" reality, i.e., of a microscopic reality *independent* of the macroscopic reality of the surrounding world. Then one hopes that it will be possible to explain in an inverted direction the macroscopic realities (from which we started the development of physical theories and which we have called $AP_i$) by this "isolated" microscopic reality. One believes that physics can begin with such a quantum mechanics describing isolated microscopic realities, and that one can *deduce* from such a theory the structure of the total world, especially the structure that is allowed to take our everyday observations as a method in which to establish realities. We do not believe in developing physics in such a way.

According to our method of physical theories we cannot use quantum mechanics (as developed in [8]) as a theory of the total world since quantum mechanics is a statistical theory over the macroscopic effects produced by microscopic realities, and we cannot have or produce a great number of separate universes. We also cannot take one universe as a reality that produces effects in a greater reality. A uniform theory of the total world cannot be a theory of a "great" number of microsystems nor a purely macroscopic theory as, e.g., the famous space–timetheory of Einstein, a theory which only describes the global interaction between the total universe and a part of this universe. Not yet solved is the problem of obtaining a detailed description of the interaction of the total world with a greater and smaller part of this world. Perhaps it is not possible to obtain such a theory, but we have to take such a theory as the intended aim of physics, and this intended theory cannot have the structure of the well-founded quantum mechanics of today.

In any case it is not necessary to believe in such fairy tales of quantum mechanics such as, e.g., "hidden variables" or "states" of individual systems, in order to understand quantum mechanics.

124    6 Real and Possible as Physical Concepts

Our criticism concerning the "too fast belief" in the reality of fairy tales does not mean that fairy tales, or some parts of fairy tales, cannot be real or cannot be very fruitful (also, if not real) for the development of physics. We are convinced that there is no method for *finding* new theories. Intuition and fairy tales can be very helpful in the development of physics. Bohr's theory of the atom as a little planetary system was a fairy tale, but nevertheless essential to the development of quantum mechanics, and the Heisenberg step to quantum mechanics consisted in not believing in some parts of this fairy tale.

For our purpose, to formulate more precisely the method of physical theories, we have to ask how we can find the real meaning of fairy tales or of parts of fairy tales. This method of finding new realities beyond $A_p$ does not consist in philosophical reflections. The only justification of such a method is its result. *After* the development of a successful method we can (and should also) make philosophical reflections. The desired method consists in the formulation of a way in $PT$ to go back from $MT_\Sigma$ to the reality.

## 6.1 Closed Theories

We will not begin to develop the intended method in a general manner. We consider at first a well-tried theory $PT \equiv A_p \leftrightarrow B_l(\text{cor})MT_\Delta \leftrightarrow MT_\Sigma$ with a given $\Delta$, i.e., with fixed $\Sigma$ and fixed inaccuracy sets for $\Sigma$. The species of structures $\Sigma$ can be a fairy tale or an axiomatic basis. We consider for such a theory at first only the problem of the connection of the elements of the terms $\overline{M}_i$ and $\overline{s}_\nu$ with the reality; the introduction of *new* concepts will be treated in Sect. 6.2.

From the "observed realities" of $A_p$ we get in $PT$ a text $\widetilde{A}$ formulated in $B_l$, and a corresponding text $\overline{A}$ formulated in $MT_\Theta$ and $MT_\Delta$. The fact that we have a well-tried theory signifies that the text $\overline{A}$ is not in contradiction to $MT_\Delta$. The text $\overline{A}$ corresponds uniquely to a realistic part of $A_p$, i.e., to that part which we have "observed." We will denote it by $W_o(\overline{A})$.

Since we have introduced the axiom that all sets $\overline{M}_i$ of $\Theta$ are finite sets, we can *imagine* having "all" real objects and "all" real relations of $A_p$ described under the form of a text $\overline{A}_{\max}$, formulated in $MT_\Theta$ and $MT_\Delta$. This signifies that $W_o(\overline{A}_{\max}) \equiv A_p$. But for all significant physical theories, we can only *imagine* having $\overline{A}_{\max}$ and $W_o(\overline{A}_{\max})$.

We will see that $W_o(\overline{A}_{\max})$ is not a given and "fixed reality" but only a "possible reality." $W_o(\overline{A})$ of a recorded text $\overline{A}$ is a fixed reality which cannot be changed; it was "observed." We cannot also expect that $\overline{A}_{\max}$ is given. On the contrary, as human beings, we are often able to change expected realities, i.e., we are able to influence future realities by our own free will. If, e.g., someone would say that something *will be* so and so, we can make it such that this something *will not be* so and so. A text $\overline{A}_{\max}$ can therefore only be one of various "possibilities." Therefore, we cannot expect that a physical theory prescribes uniquely the physical reality, it can only give us a selection of various "possible" realities. A text $\overline{A}_{\max}$ can only be one possible imagination.

There was however a time in the history of physics when one believed that all realities were determined by the so-called "initial values," i.e., that there was one and only one fixed reality forever. But this was a mistake. There are various possibilities for the development of the world. An imagined $\overline{A}_{\max}$ can only be one of the various possibilities.

Moreover, there are two reasons for which we cannot have observed one $\overline{A}_{\max}$ for almost all of the physical theories: the first reason is that for all significant physical theories, $\overline{A}_{\max}$ would be too great to be given explicitly; the second reason is that we cannot make all the necessary recordings since, e.g., we have either forgotten to do it in the past, or it would be necessary to make recordings in that region of the world that we call the future.

Therefore, as physicists, we have only the possibility of asking ourselves whether we can say something about the reality in an $A_p$ *without* having "observed" the total $A_p$. And it is perhaps the main advantage of physics, that we can indeed say something about reality, even if we have not "observed" this reality. But what is the method for obtaining assertions about a nonobserved reality?

At first one *could* perhaps think that all elements of the $\overline{M}_i$ could refer to possible real objects in the $A_p$, and the $\overline{s}_\nu$ could represent possible real relations between these objects. But this would be a premature conclusion. If we took greater sets $Q_i$ (respectively for an axiomatic basis: greater sets $\widehat{M}_i$), this could never be detected by contradictions in $MT\overline{A}$. This shows that the $Q_i$ (respectively $\widehat{M}_i$) can include elements to which there corresponds *no* reality! The $Q_i$ (respectively $\widehat{M}_i$) are "too large." How can we detect such a situation? Not by the currently used theory $PT$, but only by trying to make richer theories, e.g., in the following way.

Starting from the theory $PT$ with the picture terms $Q_i$ (terms in $MT_\Sigma$), we take as picture terms smaller $Q_{i_s} \subset Q_i$. We then try to see whether this new theory $PT_s$ is "usable" (i.e., that $\overline{A}$ does not lead to contradictions). If $PT_s$ is usable, then it also follows that $PT$ is usable. But it can be that "too small" picture terms $Q_{i_s}$ can lead to contradictions with the experiments. But how do we observe this?

The introduction of the $Q_{i_s}$ can be made by choosing intrinsic terms $Q_{i_s}$ in $MT_\Sigma$ which are subsets of $Q_i$. These terms $Q_{i_s}$ describe "mathematical qualities." For example, if we take a round table and choose $Q = \mathbb{R} \times \mathbb{R}$ (see Sect. 3.3.1), then this $Q$ is "too large." In fact, elements in $\overline{M}$, the pictures of which in $Q$ determine much greater distances than the diameter of the real table, cannot be excluded. We can then introduce $Q_s$ as the set of all $(\alpha, \beta) \in \mathbb{R} \times \mathbb{R}$ with the property $\alpha^2 + \beta^2 < r^2$, with $r$ as the measured radius of the table.

The introduction of $Q_s$ for this example is not so difficult, because we can see that "there are" no marked spots outwards from the table. In other cases in physics we cannot "survey" the existence or nonexistence of the objects corresponding to the elements of $Q$. In most cases we are in the same position as that of a blind man in front of a table. A blind man can only try to mark spots on the table and to perceive (by feeling), but he does not succeed in marking spots "outwards" from the table. Such experimental features are indications that we have "too large" picture terms $Q_i$, and that (and how) we can perhaps find smaller $Q_{i_s}$.

Another example of "too large" picture terms (where it has not been possible until now to get smaller picture terms) is the "extrapolated quantum mechanics of macrosystems."

By "extrapolated quantum mechanics," we speak of the use of quantum mechanics on macrosystems (as developed in [8]), i.e., with the *same* axioms as for microsystems. In this theory we use the picture terms $Q_1$ and $Q_2$ as "preparation procedures" and "registration procedures." These terms are "too large" for macrosystems (and also for more complicated microsystems). But how can we gain such a conviction? Not by mathematical deductions, but by reflections on how one could perhaps "realize" this. Such reflections can perhaps lead to mathematical formulations for the $Q_1$ and $Q_2$, but we have not succeeded in obtaining $Q_{1_s}$ and $Q_{2_s}$. Nevertheless, the physicists have intuitively developed very good theories for macro systems, e.g., for semiconductors.

For a gas, one can find a first approximation for $Q_{1_s}$ and $Q_{2_s}$ in [2]. The differences $Q_i \setminus Q_{i_s}$ are sets of pictures of typical "fairy objects" for which there can be no realization.

In addition to the fact that we can have "too large" picture terms $Q_i$, we can also have too many possibilities for the relations $\overline{s}_\nu$, i.e., the real possibilities of the real relations (corresponding to the $\overline{s}_\nu$) are smaller than those that are allowed by the $s_\nu$ (as the pictures of the $\overline{s}_\nu$), i.e., allowed by the axioms of $\Sigma$. The axioms of $\Sigma$ are too weak, i.e., we can strengthen these axioms without contradiction to experiments. The structure $\Sigma_s$ may be the structure $\Sigma$ with the strengthened axioms, i.e., $\Sigma_s$ is richer than $\Sigma$.

To illustrate the introduction of a richer $\Sigma_s$ without contradiction to experiments, we will briefly discuss some examples.

As a first example, we consider Newton's point mechanics. The axiom for the trajectories is Newton's axiom: $m\ddot{\vec{r}} = \vec{k}(\vec{r})$ ($m$ the mass, $\vec{r}(t)$ the trajectory, $\vec{k}(\vec{r})$ the force). All $\vec{r}(t)$ are "allowed" with various "initial values" $\vec{r}(o), \dot{\vec{r}}(o)$. We are convinced that we cannot introduce a richer $\Sigma$ without the possibility of contradictions to experiments. We can deduce in Newton's theory a theorem: the energy $\frac{1}{2}m\dot{\vec{r}}^2 + U(\vec{r})$ does not depend upon time (here $U(\vec{r})$ is given by $\vec{k}(\vec{r}) = -\mathrm{grad}\, U$). If we now define a new $\Sigma$ by replacing Newton's equation of motion by the energy theorem, we also get a usable theory. But for this $\Sigma$ we can strengthen the axioms of $\Sigma$ by Newton's equation of motion to obtain a richer $\Sigma_s$ (i.e., $\Sigma_s$ as Newton's theory).

A second example is thermodynamics without the "second law" (the entropy law). In this weaker theory, a "perpetuum mobile of the second kind" would be possible although it cannot be "realized." That it cannot be realized, cannot be "proved" by experiments; one can only be convinced of this impossibility by experimental failures to construct such a perpetuum mobile of the second kind (similar to the blind man who is convinced that marked spots "outside of the table" cannot be realized). $\Sigma_s$ with the second law is then richer than $\Sigma$ without this law.

A third example is electrodynamics with $\Sigma$ as Maxwell's equations. This $\Sigma$ would be too weak; one has to add the axiom that only retarded potentials are allowed. We are convinced of this additional axiom since we do not succeed to broadcast signals in the direction of the past.

We have used in the description of our examples words such as "possible" and "realizable." These are not words of the basic language. Thus it may seem that we have introduced these words by a loophole; but this was not our intention. On the contrary, we wanted only to describe our intention of why we define the concept of a "closed theory." *After* the introduction of this concept, we will start to define what we mean by the concepts of real and possible.

*Definition*

> A usable theory $PT$ is said to be *closed* if it is not possible to get a richer *and usable* theory $PT_s$ in one of the following ways: by introducing either smaller $Q_{i_s}$ or richer $\Sigma_s$.

The fact that we only admit these two possibilities as ways in which to obtain a richer theory $PT_s$ is based on the fact that it is not so simple to get theories that are richer in the general sense of Sect. 5.1. If we succeed in getting a richer theory in the general sense of Sect. 5.1 we have made a desirable but also very difficult progress in physics; but nobody can say in advance how such a "better" theory will look like. It is not so difficult however to get a richer $PT_s$ by following the two methods mentioned above and by making experiments to see whether such a richer theory can be usable.

*Definition*

> For a *closed* theory, we call the sets $\widehat{M}_i$ and $\widehat{s}_\nu$ of an axiomatic basis $\widehat{\Sigma}$ an idealized inaccurate picture of a *possible reality*.

(We prefer to take $\widehat{\Sigma}$ instead of $\Sigma$ with the $Q_i$ and $s_\nu$, since $\Sigma$ can contain some purely fairy tale structures that have nothing to do with reality.)

The above definition defines the use of the word "possible," relative to the reality. We mean by this word that we cannot exclude a reality for which

$\widehat{\Sigma}$ can be used as a picture. Very often one describes this situation by the following assertion: The reality obeys the so-called "laws of nature;" and a closed theory contains "all" the laws of nature for the considered region of reality (of the intended application domain). We will not use this description since it contains a certain philosophical interpretation (as, e.g., "obeys the laws of nature") and it was one of our purposes to define physics by a "method" and not by philosophical assumptions. This does not mean that, *after* using the "method," we cannot try to proceed to a philosophical interpretation. Physics is not intended to answer, e.g., the question "*why* does a stone fall to the earth?" but only the question "*how* does a stone fall to the earth?" It does not follow that it is forbidden to ask the question "why?" But physics cannot answer this question. It is essential for physics to introduce a word such as "possible" which is not based on philosophical but on methodological considerations. A method is not based on philosophy but depends only on the success of the method.

We will only give one example of the application of this new word "possible." If we use Newton's mechanics (for mass points flying around the earth) as a closed theory, we can consider trajectories of these mass points as "possible" trajectories of satellites.

But how can we decide what *is* real and not only possibly real? And how can we detect realities which cannot be formulated in the basic language $B_l$ of a theory? How can we extend the basic language in order to describe new detected realities?

## 6.2 Physical Systems

Until now we have only introduced the concept of an application domain $A_p$ as that part of reality on which the considered theory $PT$ can be applied, and therefore also as the basis on which all future concepts of indirectly observed realities can be founded. In connection with our concept of a closed theory, the concept of a "physical system" plays a central role. A physical system is a part of the application domain and therefore a part of reality that can be observed with the help of $MT$.

Often the application domain consists of various "separate" parts. We say that a part is separated if this part is not influenced by other parts of the world. The total world (the universe) is an example of such a separate part, since there is no other part which can influence the total world. We could not have developed such fruitful physics and technology if it was not possible to describe by physics much smaller parts without the various influences of the rest of the world.

If it is believed that it is impossible to separate a part of the world in such a way that it is not influenced by the rest of the world, then it would be correct since nothing can be separated *completely* from the rest of the world. But we can often describe well-defined parts as parts, e.g., a cup of tea on a table; or if we want to describe a chemical reaction in a test tube, we do not need to describe the entire laboratory. If we add milk to our cup of coffee, then it is not possible to describe that part of the "milk" without the coffee.

The application domain of a $PT$ is a part of reality that can be described by immediate observations and by pre-theories of the $PT$. It can consist of several separate parts of this kind. Our planetary system is such a part which can be described by astronomical observations and by a pre-theory of space–time; this part is the application domain of Kepler's theory (or "Kepler's laws").

The application domain will not be *influenced* by the rest of the world. The intended theory $PT$ must not take into account the processes outside of the application domain. The above-mentioned planetary system is such an application domain for Kepler's theory, since space and time can be taken as unchangeable by the rest of the world and since the influence of other astronomical objects can be neglected.

One of the essential characteristics of physical theories is that we do not need to describe the total world. Einstein's space–timetheory is the first theory by which we tried to begin with the description of global problems. The main purpose of physical theories is to try to describe as well as possible "very little parts of the world," and our technical successes are based upon these theories. We are able to do this even though it is impossible to separate precisely one part of the reality from another. But it is often possible to describe the influences of the rest of the world on the considered part, by "given" influences, i.e., by influences that do *not depend* on the influence of the considered part. Thus the "given" structure of space–timein Kepler's theory of our planetary system can be described without taking into account the influence of the planetary system on the global structure of space–time. We say that we can describe the influences of the total world by giving "boundary conditions." Also, if we want to describe the planetary system by Einstein's theory (as a more comprehensive theory than Kepler's theory), we can describe the influence of the rest of the world, e.g., by the "boundary condition" that stipulates that the space–timehas a given Euclidean structure at infinity.

*Definition*

> A part of reality is called a *physical system* if the influence of the rest of the world can be described by given boundary conditions. These boundary conditions are considered as a part of the physical system.

We will only consider in this chapter application domains that consist of a number of separate physical systems (even if there is only one system). Such application domains are very widespread. The fact that we will only consider such application domains in physical theories will not be a major restriction, since we can in any case take the total application domain as a single physical system. But taking the total application domain as a single physical system would only complicate the mathematical description in the case where the application domain consists of many separate physical systems. In the case of many separated systems, we only need to consider separately each of the separate physical systems in such a way that the application domain would be the only system. So we can take for every separate system the same $MT_\Delta$; only the observed realities are to be described by different texts $\overline{A}$. Thus we can treat each separate system separately.

Only to see more precisely the mathematical description of the total application domain, we insert here this more complicated form without using it later.

If there is more than one separate physical system, we have to formulate the relation that two objects belong to the same physical system. We do this in $\Theta$ by adding subsets $\overline{u}_{ik} \subset \overline{M}_i \times \overline{M}_k$ as structure terms: for $\overline{a} \in \overline{M}_i$ and $\overline{b} \in \overline{M}_k$ we write for "$\overline{a}, \overline{b}$ belongs to the same physical system" the relation $(\overline{a}, \overline{b}) \in \overline{u}_{ik}$. In $\widehat{\Sigma}$ we introduce corresponding sets $u_{ik} \subset M_i \times M_k$ with the axioms

$$(x, y) \in u_{ik} \Rightarrow (y, x) \in u_{ki},$$

$$(x, y) \in u_{ik} \wedge (y, z) \in u_{kl} \Rightarrow (x, z) \in u_{il}.$$

We define a subset $\overline{X}_a \subset \overline{M}_1 \cup \overline{M}_2 \cup \cdots$ by

$$\overline{X}_a = \{y \mid \exists\, \overline{u}_{ik} \,[(a, y) \in \overline{u}_{ik}]\}$$

and introduce for the $u_{ik}$ (and the corresponding $X_a$) the axiom

$$b \notin X_a \Rightarrow X_a \cap X_b = \emptyset. \tag{6.2.1}$$

We add in $\Sigma$ for the $s_\nu$ the axiom

$$(x, y, z, \ldots) \in s_\nu \Rightarrow y \in X_x \quad z \in X_x, \ldots, \tag{6.2.2}$$

i.e., $x, y, z, \ldots$ are elements of the same system.

For every system $X_a$, we postulate that the terms $M_{ia} = M_i \cap X_a, s_\nu$ form a structure of species $\widehat{\Sigma}$, independent of the system $X_a$.

There also exist much more complicated application domains where (6.2.1) is not valid. In these cases we cannot avoid considering the total application domain $A_p$.

If (6.2.1) is valid, then we can treat every system $X_a$ as an application domain with the structure $\widehat{\Sigma}$. If for the total application domain the theory is closed, then the structure $\widehat{\Sigma}$ must not be categorical, since the various systems $X_a$ cannot be isomorphic!

⌟

If we now contemplate a $\widehat{\Sigma}$ and a theory $MT_{\widehat{\Sigma}}$, this theory can only be a theory of one separate physical system.

## 6.3 New Concepts in a *PT*

Our task is not only to detect "nonmeasured" realities, but also to detect new realities which we did not know before the introduction of $\widehat{\Sigma}$. Therefore we have to seek for other concepts than those which are described by the base sets $\widehat{M_i}$ and the structure terms $\widehat{s}_\nu$ of the axiomatic basis $\widehat{\Sigma}$.

In Sect. 4.2 we have introduced the mathematical concept of intrinsic terms. These can be defined in relation to a species of structures $\Sigma$, e.g., to the axiomatic basis $\widehat{\Sigma}$. Intrinsic terms relative to $\widehat{\Sigma}$ are not new constants! They are defined by the base terms $\widehat{M_i}$, the structure terms $\widehat{s}_\nu$, and the axiomatic relations of $\widehat{\Sigma}$. They are therefore essentially different from the new constants of a fairy tale $\Sigma$ as, e.g., if one introduces in quantum mechanics a Hilbert space as a set of "states" without a definition of what a state is. Such concepts as "state" are fanciful interpretations, and no one can say whether two people will give the same meaning to the word 'state'. In contrast to such imaginary interpretations, the intrinsic terms are well-defined by the base and structure terms of $\widehat{\Sigma}$.

Since we have called the base terms and structure terms of $\widehat{\Sigma}$ for a closed theory "pictures of possible realities," we also call the intrinsic terms of $\widehat{\Sigma}$ "pictures of possible realities."

More precisely, intrinsic terms are a "deduction of structures" in $MT_{\widehat{\Sigma}}$ (see Sect. 4.2). Such a deduction is a set of intrinsic terms $E_k$ and $u_\mu$, such that the $E_k, u_\mu$ form a structure of species $\Sigma_{\text{new}}$. In a physical theory we are interested in such species of structures $\Sigma_{\text{new}}$ where the typification of the $u_\mu$ is of the particular form $u_\mu \subset E_{k_1} \times E_{k_2} \times \cdots \times \mathbb{R} \times \cdots$ (see this form in Sect. 4.1). Then it follows that for a closed theory we can say that the $E_k, u_\mu$ are pictures of newly detected "possible objects" with newly defined "possible relations." But what is meant by the word "picture"?

Before we discuss this question, we will combine such a $\Sigma_{\text{new}}$ with a $\Sigma$ which could possibly be a fairy tale structure. If the $E_k, u_\mu$ form a structure of the species $\Sigma$ and if *PT* with $MT_{\widehat{\Sigma}}$ is a closed theory, then we can say that we have proved that the fairy tale $\Sigma$ is more than just a fairy tale, i.e., that $\Sigma$ is an intuitively guessed picture of a possible reality. The fact that *PT* with $MT_\Sigma$ is a usable physical theory is *not enough* to believe that the fairy tale $\Sigma$ is a picture of a physical reality. This does not mean that in a richer theory (richer in the general sense of Sect. 5.1) the fairy tale $\Sigma$ cannot be proved to be a description of a possible reality! But before such a proof can be given in a richer theory, nobody should be forced to believe that the fairy tale is real.

We now have to analyze more precisely what is meant by "the deduced $E_k, u_\mu$ describe new detected objects and relations." To begin with, the $E_k, u_\mu$ are only terms defined mathematically. If we want to analyze the description of possible realities by the $E_k, u_\mu$, then we must go back, in $A_p \leftrightarrow B_l(\text{cor})MT_\Delta$, from $MT_{\widehat{\Sigma}}$ by $MT_\Delta$ to a reality domain $W \supset A_p$.

Let $F$ be one of the terms $E_k, u_\mu$, i.e., the term $F$ may be an intrinsic term in $MT_{\widehat{\Sigma}}$. Then $F$ is a subset of an echelon $T(\widehat{M}_1, \ldots, \mathbb{R})$ over the base sets $\widehat{M}_i$ (and eventually $\mathbb{R}$) of $\widehat{\Sigma}$. We then introduce in $MT_\Theta$ a *new constant* $\overline{F}$, finite sets $\overline{\mathbb{R}}$, and the axiom

$$\overline{F} \subset T(\overline{M}_1, \ldots, \overline{\mathbb{R}})$$

with the echelon construction scheme $T$ identical to that of $F$. This makes it thus possible to define a mapping $\phi : \overline{F} \to T(\widehat{M}_1, \ldots, \mathbb{R})$ generated in a canonical way by the mappings $\phi_i : \overline{M}_i \to \widehat{M}_i$ used under the axiom of $\Delta$ (see Sect. 3.3.1). We then take for $\Delta_{\text{ex}}$ ($\Delta$ extended) the axiom of $\Delta$ extended by the additional axiom $\phi \overline{F} \subset F_U$, where $F_U$ is the set $F$ increased by the inaccuracy sets (see Sect. 4.6). Such an axiom $\Delta_{\text{ex}}$ is not a new condition for the results of experiments! It is a condition for the new constant $\overline{F}$ in $\Delta_{\text{ex}}$ and it says that $\overline{F}$ is approximated by the "idealization" $F$. The question of knowing which size the inaccuracy set $U$ of $F$ must take will be discussed in Sect. 6.4.

After the introduction of $\overline{F}$ we have to invent a new word (designating a new concept) for this new possible reality described by $\overline{F}$. By adding this new word to $B_l$ we get an extended language $B_{\text{lex}}$. We can then speak about this new possible reality in the same way as we can speak about reality with the words of $B_l$. In particular, we call the new concepts introduced for the $\overline{E}_k, \overline{u}_\mu$ (the new constants in $\Delta_{\text{ex}}$ for the intrinsic terms $E_k, u_\mu$ introduced above) "classification concepts" for the $\overline{E}_k$ and "relation concepts" for the $\overline{u}_\mu$ respectively. Such new classification concepts are, e.g., "electron," "H-atom," "neutron," "proton" but also "temperature" and "entropy."

It can be that some $E_k$ are sets $\widehat{M}_i$ (or subsets of the $\widehat{M}_i$). Nevertheless, $\overline{E}_k$ is a different constant than $\widehat{M}_i$! In this way we can also get in $B_{\text{lex}}$ different words for $\overline{E}_k$ than for $\overline{M}_i$, even if, e.g., the "pictures" $\widehat{E}_k$ and $\widehat{M}_i$ of the different $\overline{E}_k$ and $\overline{M}_i$ are the same. This sometimes leads to misunderstandings if, e.g., "preparation procedures" and "micro-objects" such as electrons have the same "picture."

By the introduction of new words (designating new concepts) in $B_{\text{lex}}$, we have only said how we speak about these new possible realities. We have not said how we can "observe" such new realities. To observe the "basic properties" (corresponding to the $\overline{M}_i$) and the "relations" (corresponding to the $\overline{s}_\nu$) we could use immediate observations and "pre-theories," i.e., only well-defined methods before the introduction of the theory $MT$. Now we have to discuss how we can obtain from the text $\widetilde{A}$ an extended text $\widetilde{A}_{ex}$, where in $\widetilde{A}_{ex}$ are also used the words of $B_{lex}$. We have to discuss what we mean, e.g., by a sentence of $B_{lex}$ such as 'we have observed an electron'.

This problem is much more difficult than the introduction of new physical concepts. We will begin with an analysis of the problem of the "observation" of new concepts in Sect. 6.4.

We finish this section by giving a simple example for the introduction of new concepts. We take for this purpose our

*Example A*

For a round table, $\widehat{\Sigma}$ is given by the set $M$ of points in a circle of radius $r$ (the table has a radius of $r \pm \varepsilon$), a structure term $s \subset M \times M \times \mathbb{R}$, and the axiomatic relation that $s$ determines a distance $d : M \times M \to \mathbb{R}$, so that $M$ is the set of points in a circle of radius $r$ of a two-dimensional Euclidean space. This theory of the round table is a closed theory. ($\widehat{\Sigma}$ is in the mathematical sense not an Euclidean geometry!)

As a new intrinsic term $E_2$ we introduce the subset $E_2 \subset \mathcal{P}(M)$ of "straight lines" (lying *in* the circle). As $E_1$ we take $E_1 = M$. As a new relation $u \subset E_1 \times E_2$ we take the set $u = \{(x,y) \mid x \in y, x \in E_1, y \in E_2\}$ (i.e., "$x$ is a point on the straight line $y$"). To get $MT_{\Delta_{ex}}$, we add to $\Theta$ the new constant $\overline{E}_2$ with the axiom $\overline{E}_2 \subset \mathcal{P}(\overline{M})$, a relation $\overline{u} \subset \overline{M} \times \overline{E}_2$, and the axiomatic relation that the mapping $\phi : \overline{M} \to M$, of the axiomatic relation of $\Delta$, fulfills the relations $\phi \overline{E}_2 \subset (E_2)_U$ and $\phi \overline{u} \subset (u)_U$, where $(E_2)_U, (u)_U$ are the "inaccurate" $E_2$ and $u$, i.e., $E_2$ and $u$ enlarged by the inaccuracy sets $U$.

As new sentences (expressing new propositions) in $B_{lex}$ we introduce the following correspondences:

for '$\overline{y} \in \overline{E}_2$' : '$\overline{y}$ is a marked straight line';

for '$(\overline{x}, \overline{y}) \subset \overline{u}$' : 'the marked spot $\overline{x}$ lies on the marked straight line $\overline{y}$'.

The fact that a marked straight line is determined by two marked spots lying on this marked line has to be proved. How this can be done will be shown in Sect. 6.4.

## 6.4 Indirect Measurements

If we have detected a trace in a cloud chamber, we can, e.g., conclude that this trace was produced by an "electron," i.e., that we have established the reality of an "electron." To make such a conclusion we must have a theory $PT$ where the trace belongs to a text $\widetilde{A}$, and with the help of which from the reality described by $\widetilde{A}$ we can infer the "reality of an electron." It is our aim to provide the *method* by which we can make such conclusions and, like every method, it can only be justified by the success we obtain from working with it!

To begin with the development of such a method, we will at first describe the general form of such conclusions.

We have, as described in Sect. 6.3, a theory $PT \equiv A_p \leftrightarrow B_l(\text{cor})MT_\Delta$ which is extended to $PT_{\text{ex}} \equiv A_p \leftrightarrow B_{l_{\text{ex}}}(\text{cor})MT_{\Delta_{\text{ex}}}$, where $\Delta_{\text{ex}}$ has the base terms $\overline{M}_i$ and $\overline{E}_k$, the structure terms $\overline{s}_\nu$ and $\overline{u}_\mu$, and where the mapping $\phi$ of the axiomatic relation of $\Delta_{\text{ex}}$ maps $\overline{M}_i \to M_i$, $\overline{s}_\nu \to (s_\nu)_U$, $\overline{E}_k \to (E_k)_U$, and $\overline{u}_\mu \to (u_\mu)_U$. Here, the $E_k, u_\mu$ are intrinsic terms of $MT_{\widehat{\Sigma}}$ by which a structure $u_\mu$ over the $E_k$ of species $\Sigma_{\text{new}}$ is "deduced" from $\widehat{\Sigma}$.

We have to develop a method by which from a text $\overline{A}$ (the result of "direct measurements") we can infer the "reality" of an extended text in which are also used the $\overline{E}_k$ and $\overline{u}_\mu$ in the same manner that in $\overline{A}$ are used the $\overline{M}_i$ and $\overline{s}_\nu$, i.e., a method by which we can "interpret" the directly observed text $\overline{A}$ as an "indirect measurement" of a "hypothetical" text in which the $\overline{E}_k$ and $\overline{u}_\mu$ also appear.

Such an extended text consists of two parts:

(i) The directly measured part $\overline{A}$ composed of relations of the form

$$\overline{A} : \overline{a}_1 \in \overline{M}_i, \ldots;$$
$$(\overline{a}_1, \overline{a}_2, \ldots, J_1, J_2, \ldots) \cap \overline{s}_\nu \neq \emptyset, \ldots \qquad (6.4.1)$$

(the $J_l$ are intervals of $\overline{\mathbb{R}}$, a finite set of real numbers),

(ii) The "hypothetical" part $\mathcal{H}$ composed of relations of the form

$$\mathcal{H} : \overline{y}_1 \in \overline{M}_i, \ldots \qquad (6.4.2)$$
$$(\overline{y}_1, \ldots, \overline{a}_1, \ldots, \eta_i \ldots, J_k \ldots) \cap \overline{s}_\nu \neq \emptyset, \ldots \qquad (6.4.3)$$
$$\overline{x}_1 \in \overline{E}_k, \ldots \qquad (6.4.4)$$
$$(\overline{a}_1, \overline{x}_2, \overline{y}_3, \eta_i, J_k, J_l, \ldots) \cap \overline{E}_l \neq \emptyset, \ldots \qquad (6.4.5)$$
$$(\overline{x}_1, \overline{x}_2, \ldots, \gamma_1, \gamma_2, \ldots) \in \overline{U}_\mu, \ldots \qquad (6.4.6)$$
$$(\overline{x}_1, \ldots, \overline{a}_1, \ldots, \overline{y}_1, \ldots, \eta_i, J_k, J_l, \gamma_1, \ldots) \cap \overline{u}_\nu \neq \emptyset, \ldots \qquad (6.4.7)$$

The text $\overline{A}$ describes the stated facts by giving to the stated objects signs $\overline{a}_\nu$, and by writing down stated relations by $(\overline{a}_1, \overline{a}_2, \ldots, J_1, J_2, \ldots) \cap \overline{s}_\nu \neq \emptyset$. The $J_\nu$ are given intervals, given by the experiments.

The text $\mathcal{H}$ describes hypothetical objects $\overline{x}_\nu, \overline{y}_\nu$, i.e., the text $\mathcal{H}$ does not say that a reality belongs to a particular letter $\overline{x}_\nu$ or $\overline{y}_\nu$. The text $\mathcal{H}$ also describes hypothetical relations between these hypothetical objects $\overline{x}_\nu, \overline{y}_\nu$, the stated objects $\overline{a}_\nu$, real numbers (described by hypothetical real numbers

$\eta_i, \gamma_k$), and intervals $J_\nu, J_k$ of real numbers. The $J_\nu$ are intervals of numbers $\eta_i$ stated by experiments written in $\overline{A}$. The $J_k$ are intervals of numbers $\eta_i$, intervals specially selected and hypothetically selected. The $\gamma_\nu$ are hypothetical numbers.

The fact that we have used the numbers $\eta_i$ which describe objects already known by the pre-theories (e.g., time and space by Newton's pre-theory) in two ways, either as a hypothetical number $\eta_i$ or as a hypothetical interval $J_k$, has to do with the intention of what we want to describe by $\overline{\mathcal{H}}$.

The numbers $\eta_i$ describe *unknown* values of hypothetical objects. The intervals $J_k$ describe *fixed* regions of realities or possible realities. We hope to see how the text $\overline{A}$ can tell us which of the $\eta_i$ are allowed, i.e., how the $\eta_i$ can be measured indirectly by $\overline{A}$. The intervals $J_k$ are "given" and $\overline{A}$ can only say whether such a given interval is real or at least possibly real, or is wrong.

(How real numbers appear in physical theories is described in Sect. 4.8. A known form is, e.g., to use real numbers to describe time and to use triple of real numbers to describe spots in space; this description presumes that we have fixed by pre-theories and experiments a so-called "reference system" of space–time.)

Relations of the form (6.4.5) and (6.4.7) are possible if the $\overline{E}_l$ are sets or subsets of sets of the form $\overline{M}_{i_1} \times \overline{M}_{i_2} \times \cdots \mathbb{R} \times \cdots$. It is essential that we consider relations that connect hypothetical objects with stated objects in the form of (6.4.5) and (6.4.7).

(In (6.4.1) to (6.4.7) we did not write "possible negations" of relations. We did this because such negations do not provide us with any new points of view.)

The main question is whether and how we can state the relations (6.4.2) to (6.4.7) from the relation (6.4.1) (describing the already stated facts) with the help of the theory $MT_{\widehat{\Sigma}_{\text{ex}}}$. If we have succeeded in proving that the hypothetical objects $\overline{x}_\nu, \overline{y}_\mu$ and the hypothetical numbers $\eta_i, \gamma_k$ are almost determined by $\overline{A}$, then we say that we have indirectly stated the objects $\overline{x}_\nu, \overline{y}_\mu$, the number $\eta_i, \gamma_k$, and the relations (6.4.3), (6.4.6), and (6.4.7), i.e., that by the direct measurement $\overline{A}$ we have indirectly measured $\overline{\mathcal{H}}$.

To discuss this main question we transport it from $MT_{\widehat{\Sigma}_{\text{ex}}}$ to $MT_{\widehat{\Delta}}$ by a mapping $\phi$ which exists according to the axiomatic relation for $\Delta_{\text{ex}}$ (see Sect. 6.3) and the definition of the $\overline{E}_k$ (given in Sect. 6.3). With $\phi \overline{a}_\nu = a_\nu$, $\phi \overline{x}_\nu = x_\nu$, and $\phi \overline{y}_\nu = y_\nu$ we get from (6.4.1) to (6.4.7) the text

$$A : a_1 \in M_i, \ldots \qquad (6.4.8)$$

$$(a_1, a_2, \ldots, J_1, J_2, \ldots) \cap (s_\nu)_U \neq \emptyset, \ldots \qquad (6.4.9)$$

$$\mathcal{H} : y_1 \in M_i, \ldots \qquad (6.4.10)$$

$$(y_1, \ldots, a_1, \ldots, \eta_i \ldots, J_k \ldots) \cap (s_\nu)_U \neq \emptyset, \ldots \qquad (6.4.11)$$

$$x_1 \in (E_k)_U, \ldots \qquad (6.4.12)$$

$$(a_1, x_2, y_3, \eta_i, J_k, J_l, \ldots) \cap (E_l)_U \neq \emptyset, \ \ldots \quad (6.4.13)$$

$$(x_1, x_2, \ldots, \gamma_1, \gamma_2, \ldots) \in (u_\mu)_U, \ \ldots \quad (6.4.14)$$

$$(x_1, \ldots, a_1, \ldots, y_1, \ldots, \eta_i, J_k, J_l, \gamma_1, \ldots) \cap (u_\nu)_U \neq \emptyset \ . \quad (6.4.15)$$

If we have in $\mathcal{H}$ only relations of the form (6.4.10), (6.4.11), we call $\mathcal{H}$ a hypothesis of the first kind where the $J_k$ are given and fixed. Our question is then whether from the direct measurements (6.4.8), (6.4.9) we can calculate the results of possible measurements of the form (6.4.10), (6.4.11). This is a well-known procedure in physics: to calculate, from direct measurement, not directly measured realities. For instance of a round table, to calculate, on the basis of the direct measurement of distances between certain spots, distances not directly measured between certain spots.

For the general question of how a text $A$ can be considered as an indirect measurement of $\mathcal{H}$, we begin with the discussion of a mathematical problem: How $A$ can determine, or at least restrict, the mathematical possibilities of a hypothesis $\mathcal{H}$. To analyze how $A$ restricts $\mathcal{H}$, we will change, i.e., simplify, the form of the text (6.4.8) to (6.4.15).

We introduce the following abbreviations:

$$B \equiv (a_1, a_2, \ldots), \quad Y \equiv (y_1, y_2, \ldots), \quad X \equiv (x_1, x_2, \ldots), \quad (6.4.16)$$

and

$$D \equiv (B, Y), \quad Z \equiv (X, Y). \quad (6.4.17)$$

The relations $a_1 \in M_1, \ldots$ of (6.4.8) can be taken together in the form

$$B \in T_r(M_1, \ldots) \quad (6.4.18)$$

with

$$T_r(M_1, \ldots) = M_1 \times \cdots \quad (6.4.19)$$

The relations (6.4.10) can be taken together in the form

$$Y \in T_h(M_1, \ldots), \quad (6.4.20)$$

where $T_h$ is an echelon construction scheme similar to $T_r$. Since the $E_k$ are subsets of echelon sets of the $M_i$ and $\mathbb{R}$, the relations (6.4.12) with an intrinsic set $E$ can be taken together in the form

$$X \in E. \quad (6.4.21)$$

We can then write $MT_{\widehat{\Sigma}}A$ as the theory with the constants $M_i$, the constant $B$, and the axiomatic relation

$$P_A \equiv \widehat{P} \wedge B \in T_r(M_1,\ldots) \wedge (6.4.9). \tag{6.4.22}$$

In this theory $MT_{\widehat{\Sigma}}A$, we consider $\mathcal{H}$ as the relation

$$\mathcal{H} \equiv X \in E \wedge Y \in T_h(M_1,\ldots) \wedge Q(X,Y,\Gamma,H), \tag{6.4.23}$$

where $Q$ is the relation

$$Q \equiv (6.4.11) \wedge (6.4.13) \wedge (6.4.14) \wedge (6.4.15), \tag{6.4.24}$$

and where $H$ (majuscule letter of $\eta$!) is the row of the real numbers $(\eta_1, \eta_2, \ldots)$, and $\Gamma$ the row of the $(\gamma_1, \gamma_2, \ldots)$.

With $Z$ of (6.4.17) we can write (6.4.23) in the form

$$\mathcal{H} \equiv Z \in E \times T_h(M_1,\ldots) \wedge Q(Z,\Gamma,H). \tag{6.4.25}$$

By the relation (6.4.25) is defined in $MT_{\widehat{\Sigma}}A$ the intrinsic term

$$G(A,\mathcal{H}) \equiv \{(Z,\Gamma,H) \mid \mathcal{H}\} \subset E \times T_h(M_1,\ldots) \times \mathbb{R} \times \mathbb{R} \times \cdots, \tag{6.4.26}$$

where $\mathbb{R} \times \mathbb{R} \times \cdots$ contains as many factors as the rows $\Gamma$ plus $H$. The hypothesis $\mathcal{H}$ can then also be written in the form

$$\mathcal{H} \equiv (Z,\Gamma,H) \in G(A,\mathcal{H}). \tag{6.4.27}$$

Our problem is to analyze how $A$ and the fixed intervals $J_k$ of $\mathcal{H}$ determine the set $G(A,\mathcal{H})$. At first, we assert that

By the "direct measurement" $A$ (the "experiment" $A$) and the fixed $J_k$, we "indirectly measure" the hypothesis $\mathcal{H}$ with the "inaccuracy" $G(A,\mathcal{H})$.

It is clear that new detected realities cannot be "founded" on such an assertion. But $G(A,\mathcal{H})$ in (6.4.27) is the short description of the work of experimental physicists, i.e., of "indirect measurements." If $G(A,\mathcal{H})$ is very "large," the result of the indirect measurement says practically nothing about the reality.

But how we interpret an indirect measurement will be treated in Sect. 6.5.

Before that we will briefly formulate an often used relation that describes the seeking of "good" experiments. In this way we do not ask for the calculations of indirect measurements of relations of the form (6.4.10), (6.4.11). We want to use these relations for the seeking of direct measurements that are suitable for obtaining "good" indirect measurements.

Therefore we now take together the relations (6.4.8) to (6.4.11) as a relation $A_h$, and the relations (6.4.12) to (6.4.15) as a relation $\mathcal{H}_h$. We then use $A_h$ (for fixed numbers $\eta_i$) in the same way as the relation $A$ above, i.e., we consider the theory $MT_{\widehat{\Sigma}}A_h$ instead of $MT_{\widehat{\Sigma}}A$ and define the set

$$G_h(A_h, \mathcal{H}_h) \equiv \{(X, \Gamma) \mid \mathcal{H}_h\} \subset E \times \mathbb{R} \times \mathbb{R} \times \cdots \qquad (6.4.28)$$

and seek for such $A_h$ that $G_h$ is not too large.

## 6.5 Classifications and Interpretations

We begin our analysis with a simple case: a hypothesis of the first kind. We have already defined this case in Sect. 6.4. $\mathcal{H}$ is of the first kind if it contains only relations of the form (6.4.10), (6.4.11). Then there is no $X$ (therefore $Z = Y$) and no $\Gamma$. The set $G_h(A, \mathcal{H})$ takes the form

$$G_h(A, \mathcal{H}) \equiv \{(Y, H) \mid \mathcal{H}\} \subset T_h(M_1, \ldots) \times \mathbb{R} \times \mathbb{R} \times \cdots , \qquad (6.5.1)$$

where $\mathbb{R} \times \mathbb{R} \times$ corresponds to the set $H$ of real numbers.

If there is also no $Y$, we have the particular case that $\mathcal{H}$ contains no hypothetical objects, but only hypothetical relations, i.e., not directly measured relations! Thus $G_h(A, \mathcal{H})$ is only a subset of $\mathbb{R} \times \mathbb{R} \times \ldots$, i.e., a set $H$ of real numbers for which $\mathcal{H}$ is valid.

If there is also no $H$, then there is no set $G_h(A, \mathcal{H})$ but only a relation $\mathcal{H}$ of hypothetical relations between the directly observed objects $a_i$. For this simplest case we also have the simplest classification:

(a)  $\mathcal{H}$ is a theorem in $MT_{\widehat{\Sigma}}A$,

(b)  $\neg\mathcal{H}$ is a theorem in $MT_{\widehat{\Sigma}}A$, $\qquad\qquad$ (6.5.2)

(c)  $\mathcal{H}$ and $\neg\mathcal{H}$ can be added to $MT_{\widehat{\Sigma}}A$ without contradiction.

We have the following equivalences:

(a) is equivalent to "$A \Rightarrow \mathcal{H}$ is a theorem in $MT_{\widehat{\Sigma}}$,"

(b) is equivalent to "$A \Rightarrow \neg\mathcal{H}$ is a theorem in $MT_{\widehat{\Sigma}}$,"

(c) is equivalent to "$A \wedge \mathcal{H}$ and $A \wedge \neg\mathcal{H}$ can be added to $MT_{\widehat{\Sigma}}$ without contradiction."

The cases (a), (b), and (c) are only statements of a "mathematical classification." We also introduce a "physical classification" which is based on experiments. Therefore the cases of this classification cannot be "proved" in the same way as the cases of the mathematical classification. We introduce the following physical classification:

## 6.5 Classifications and Interpretations

($\alpha$) $\mathcal{H}$ can be added to $MT_{\widehat{\Sigma}}A_{\text{ex}}$ without contradiction, where $A_{\text{ex}}$ is *any real* extension of $A$, i.e., a description of realities that include $A$;

($\beta$) there is an extension $A_{\text{ex}}$ which is in contradiction to $\mathcal{H}$.

(6.5.3)

We also say in the case ($\alpha$) that "$\mathcal{H}$ is certain," and in the case ($\beta$) that "$\mathcal{H}$ is uncertain."

For a usable physical theory $PT$, for every $A_{\text{ex}}$ the theory $MT_{\widehat{\Sigma}}A_{\text{ex}}$ is without contradiction, therefore:

– in the case (a) $\mathcal{H}$ is certain,
– in the case (b) $\mathcal{H}$ is uncertain,
– in the case (c) $\mathcal{H}$ can be both certain and uncertain.

If in the case (c) $\mathcal{H}$ is certain, we can then add to $MT_{\widehat{\Sigma}}A$ the relation $\mathcal{H}$ without contradiction to all experiments, i.e., by adding to $MT_{\widehat{\Sigma}}$ the relation $A \Rightarrow \mathcal{H}$, we get a more comprehensive theory which is also usable. If the theory is closed (see Sect. 6.1), by adding to $MT_{\widehat{\Sigma}}$ the relation $A \Rightarrow \mathcal{H}$, we should not get a more comprehensive theory; that is, a $\mathcal{H}$ certain should also fulfill (a), i.e., in the case (c) $\mathcal{H}$ should be uncertain.

Thus for a closed theory, we have that "(a) and ($\alpha$) are equivalent" and "from (b) and from (c) follows ($\beta$)."

We make the following interpretations:

– in the case (a) we say that "$\mathcal{H}$ describes a reality," briefly that "$\mathcal{H}$ is real;"
– in the case (b) we say that "$\mathcal{H}$ does not describe a reality," briefly that "$\mathcal{H}$ is unreal" or "$\mathcal{H}$ is wrong;"
– in the case (c) we say that "$\mathcal{H}$ is both possibly real and possibly unreal."

For these classifications and the following interpretations one could perhaps wish to have something like a "proof" or a "philosophical foundation." But such a foundation is *not given* by the method of physics as described here. As in the simplest case, if we say that "a stone falls to the earth," we have no other foundation than that it is very likely that the next stone will also fall to the earth, although we do not know "why" the stone falls to the earth. Physics describes the reality but *does not say why* the reality is such as physics describes it.

We must now extend the above interpretation of $\mathcal{H}$ to more and more complicated cases. At first we take the above described case of a hypothesis of the first kind, where $G(A, \mathcal{H})$ is a subset of $\mathbb{R} \times \mathbb{R} \times \cdots$, i.e., where there is no $Y$ in (6.5.1). Then we can have the following classification:

(a) $G(A, \mathcal{H}) \neq \emptyset$ is a theorem in $MT_{\widehat{\Sigma}}A$,

(b) $G(A, \mathcal{H}) = \emptyset$ is a theorem in $MT_{\widehat{\Sigma}}A$,

(6.5.4)

(c) $G(A, \mathcal{H}) \neq \emptyset$ and $G(A, \mathcal{H}) = \emptyset$ can be added to $MT_{\widehat{\Sigma}}A$ without contradiction.

Similar to the above, we introduce the following classification:

($\alpha$) $G(A, \mathcal{H})$ is "certain" if $G(A, \mathcal{H}) \neq \emptyset$ can be added to $MT_{\widehat{\Sigma}} A_{\text{ex}}$ without contradiction, where $A_{\text{ex}}$ is any real extension of $A$;

(6.5.5)

($\beta$) $G(A, \mathcal{H})$ is "uncertain" if there is an extension $A_{\text{ex}}$ for which "$G(A, \mathcal{H}) \neq \emptyset$" cannot be added to $MT_{\widehat{\Sigma}} A_{\text{ex}}$ without contradiction.

For a closed theory, we expect that "(a) and ($\alpha$) are equivalent."
We make the following interpretations:

– in the case (a) we say that "$\mathcal{H}$ is real with an inaccuracy $G(A, \mathcal{H})$;" or that "$\mathcal{H}$ has been indirectly measured, by the direct measurement $A$, with an inaccuracy $G(A, \mathcal{H})$;"
– in the case (b) we say that "$\mathcal{H}$ is unreal for any $H$;"
– in the case (c) we say that "$\mathcal{H}$ is possibly unreal."

Most interesting is the case (a), where we get information about the reality without directly measuring all of the relations. We can obtain such information because we know that the theory $MT_{\widehat{\Sigma}}$ is a description (an imprecise one) of the reality. For example, we must not directly measure all the distances between the marked spots $a_i$; the measurement of some distances allows us to "calculate" (with a certain imprecision) the other distances, because we know the (imprecise) Euclidean structure of the surface of the table.

For measuring distances, in general one uses (as is well known) a theory that comprises not only a theory of measurement of distances by measuring tapes, but also a theory that comprises a geometric optic which allows us to calculate indirectly many not directly (by a measuring tape) measured distances.

The indirect measurements are so widespread in physics that such work is made without thinking about their theoretical base. A lot of effort is made in order to obtain a high measuring accuracy. But why? An imprecise measurement is also a reality, but it does not say as much as desired about the intended reality. We have therefore said above for the case (a) that $\mathcal{H}$ describes a reality (i.e., the intended reality) with an inaccuracy $G(A, \mathcal{H})$. If $G(A, \mathcal{H})$ is too "large," then we do not get as much information about the reality as "desired." One of the essential works of experimental physicists is to seek for indirect methods of measurement.

If we now go over the general form of a hypothesis of the first kind, we have to interpret $G(A, \mathcal{H})$ of the general form (6.5.1). At first we have the classification as in (6.5.4) and (6.5.5); we have only to say how we interpret $Y$ and $H$, the hypothetical objects $y_i$, and the hypothetical indirect measurement results $\eta_i$.

We make the following interpretations:

- In the case (a) we say that $(Y, H) \in G(A, \mathcal{H})$ are the results of indirect measurements, i.e., the $y_i$ as indirectly "recorded" objects and the $\eta_k$ as indirectly measured results, recorded and measured with the inaccuracy $G(A, \mathcal{H})$. In this sense $G(A, \mathcal{H})$ is an inaccurate description of a reality. The description of the reality becomes better and better if $G(A, \mathcal{H}) \neq \emptyset$ becomes smaller and smaller. If $G(A, \mathcal{H})$ is very large, then we have practically no description of the reality. The accuracy depends upon $A$.
- in the case (b) we say that "$\mathcal{H}$ is unreal;"
- in the case (c) we say that "$\mathcal{H}$ is both possibly real and possibly unreal;" only other experiments than $A$ can decide whether $\mathcal{H}$ is real or not.

On the basis of the mapping $\phi_i : \overline{M}_i \to \widehat{M}_i$ (see Sect. 6.3) we can go back to the hypothesis $\overline{\mathcal{H}}$ (see Sect. 6.4) and introduce a set $\overline{G}(\overline{A}, \overline{\mathcal{H}})$ as the set of all elements which are mapped onto $G(A, \mathcal{H})$. Thus we can transport the interpretations onto $\overline{\mathcal{H}}$:

- in the case (a) we say that "$\overline{\mathcal{H}}$ is real with an inaccuracy $\overline{G}(\overline{A}, \overline{\mathcal{H}})$,"
- in the case (b) we say that "$\overline{\mathcal{H}}$ is unreal,"
- in the case (c) we say that "$\overline{\mathcal{H}}$ is both possibly real and possibly unreal."

It is not now difficult to transport all of these interpretations to the most general case of a hypothesis $\overline{\mathcal{H}}$ or $\mathcal{H}$. We will not repeat all of these interpretations. We add only the often used word for a hypothesis $\mathcal{H}$ in the case (a) that we have "detected" the "new" objects $x_i$ (resp. $\overline{x}_i$) instead of the also used word "recorded." In such a way we can show that we have detected, e.g., an "electron," a "proton," or a "He-atom" with the help of an axiomatic basis for quantum mechanics. To prove in the same way that the "quarks" are real has not been established until now because it appears that we have not finished finding an axiomatic basis.

To illustrate the mathematical working of the introduction of new concepts, we take again a round table. As $MT_{\widehat{\Sigma}}$ we have a circle of a two-dimensional Euclidean geometry described by a set $M$ of points and a distance relation $s(x, y, \eta) \subset M \times M \times \mathbb{R}$ (see Sect. 4.3). As $E_1, E_2$ we introduce

$$E_1 = M, \quad E_2 = \text{the set of straight lines}. \tag{6.5.6}$$

As relation $u$ we introduce

$$u = \{(y, x) \mid y \in E_1, x \in E_2, y \in x\}. \tag{6.5.7}$$

As given recorded $A$ we assume

$$A: \ a_1 \in M \ , \ a_2 \in M \ , \ d(a_1, a_2) \sim \eta, \tag{6.5.8}$$

where $d(a_1, a_2) \sim \eta$ is briefly written for the measurement of the distance between the two marked spots $a_1, a_2$ as appropriate $\eta$.

142   6 Real and Possible as Physical Concepts

As $\mathcal{H}$ we introduce

$$\mathcal{H}\colon x \in (E_2)_U,$$
$$a_1 \in E_1 \;,\; a_2 \in E_2, \qquad\qquad (6.5.9)$$
$$(a_1, x) \in (u)_U \;,\; (a_2, x) \in (u)_U,$$

that is, the recorded spots $a_1, a_2$ lie approximately on the line $x$.

The set $G(A, \mathcal{H})$ is then the set of all straight lines $x$ for which $a_1$ and $a_2$ lie approximately on these lines. We see that $G(A, \mathcal{H}) \neq \emptyset$ is a theorem. If $d(a_1, a_2)$ is large compared to the inaccuracy set $U$, the set $G(A, \mathcal{H})$ is only a neighborhood of one straight line. If $d(a_1, a_2)$ is smaller than $U$, the set $G(A, \mathcal{H})$ contains a total cluster of straight lines around the two spots $a_1, a_2$ lying together. We then say that the two spots $a_1, a_2$ record a straight line if they are sufficiently separated.

Later one can state by experiment that a "ruler" can also be used to record a straight line (e.g., by drawing with the ruler and a pencil a straight line onto the round table) which is defined as being an element of $E_2$.

This example should only demonstrate a little bit the significance of $G(A, \mathcal{H})$ as an inaccuracy of an indirect measurement. The aim of experimental physics is to make this inaccuracy $G(A, \mathcal{H})$ as small as possible by changing the apparatuses, i.e., by obtaining an $A$ which makes $G(A, \mathcal{H})$ as small as possible. For this purpose we consider the set $G_h(A_h, \mathcal{H}_h)$ of (6.4.28) and have to seek for such new experiments, i.e., new realizations of theoretical possibilities, so that $G_h(A_h, \mathcal{H}_h)$ is small. The intention of this aim is to make new and better experiments. New and better theories is the intention of theoretical physics.

But there is also another intention: the realization of possibilities for solving the problems of our lives. This so-called "technique" has developed many new methods which have been used to discuss the possibilities of realizations of hypotheses $\mathcal{H}$; these methods are so many that it is not possible to survey all of them (which are applied by so-called "specialists").

At the end of this chapter, we will emphasize again that the interpretation "$\mathcal{H}$ is possibly real" is only meaningful if we have a closed theory. Sometimes errors are made if we neglect this condition. As an example we will mention the theories of many atoms as theories of macroscopic bodies.

If we take, e.g., quantum mechanics to describe the motion of tennis balls, we can come to the conclusion that this theory will not be closed. Such a conclusion is not based on a mathematical proof; it is an opinion based on physical experiments, but not of the sort that $\overline{A}$ is in contradiction with $MT_\Delta$. One obtains no contradictions between $\overline{A}$ and the theory, but only insuperable difficulties in realizing a "possibility," as, e.g., that it seems "impossible" to make similar diffraction experiments with tennis balls as, e.g., with electrons or neutrons.

Or, if we take the quantum mechanics of many atoms as a theory of gas, this theory is not in contradiction with experiment, but not closed. A motion of a gas is irreversible, i.e., a time-reversed motion to a real motion cannot be realized, and this cannot be described by quantum mechanics of many atoms.

One can try going further than the quantum mechanics of many atoms to a closed theory by adding additional axioms for the "realizable" preparation procedures and "realizable" registration procedures. But we are far away from a closed quantum theory of macroscopic bodies. Quantum mechanics cannot be used for the total cosmos since we have no preparation procedures for many cosmos to imply statistics, and no registration procedures to record outside of the cosmos. We only have one cosmos.

The realizable preparation procedures for macroscopic bodies (small compared to the cosmos) are determined by the given cosmos. The irreversibility produced by these realizable preparation procedures seems to be, in this way, a consequence of the expanding structure of the cosmos. Also, the emission of light in the direction of time, where the cosmos is expanding, seems not to be a mere accident.

It still remains an unsolved problem to find a closed theory for macro systems. Nevertheless, it is astonishing how theoretical physicists can intuitively find realizable preparation procedures for many special cases such as, e.g., for a theory of semiconductors.

## 6.6 The Reality Domain of a *PT*

In the previous paragraph we have given an interpretation of the set $G(A, \mathcal{H})$ relative to the hypothesis $\mathcal{H}$, and have especially said what we mean by "$\mathcal{H}$ is real." Now we will try to give a survey of the totality of reality which can be described by a theory *PT*, and also to look at realities that cannot be described by physics.

We began in Sect. 3.1.2 with the development of a *PT* by introducing the application domain $A_p$, as the domain of realities which can be stated without any theory or with the help of pre-theories. Thus, stated facts can be described by a text $\widetilde{A}$ formulated in the basic language $B_l$. The text $\widetilde{A}$ is a description of a part of $A_p$. We can also briefly say (but not very correctly!) that "$\widetilde{A}$ is a part of $A_p$." In this way we "identify" the text $\widetilde{A}$ with a part of the reality.

The text $\widetilde{A}$ is in this sense not "all" that we know of $A_p$, and never all of $A_p$. If we take together all that we know (which is practically not possible), we can write for this special $\widetilde{A}$: $\widetilde{A}_{\text{tot}}$. The text $\widetilde{A}_{\text{tot}}$ is in this sense the totality of all *stated* facts of $A_p$. The sense of a *PT* is not only that $\widetilde{A}_{\text{tot}}$ is not in contradiction with the mathematical part $MT_\Delta$ of *PT*; the sense is mostly that we can detect and construct *new* realities, new realities in $A_p$ which are not stated by $\widetilde{A}_{\text{tot}}$, but also new realities as real hypotheses $\mathcal{H}$ described by new concepts in the sense of Sect. 6.3.

The totality of all these realities, i.e., the reality domain of all $PT$s (denoted by $W$) was introduced in Sect. 1.2. The reality domain of a particular $PT_\nu$ (denoted by $W_\nu$) is in this sense a part of $W$ (the reality domain $W$ is not a fixed reality). We can form a part of $W$ by realizing "possible" hypotheses $\mathcal{H}$. The way in which to realize a possible hypothesis can be very complicated, so that a realization is only possible by working as a team. But there also arises a new problem as to what we "can" realize and also as to what we will realize and we should "not" realize. By physics we can only find what we can realize, but not what we will realize and what we should not realize.

The reality domain $W_\nu$ of a $PT_\nu$ increases with the work of physicists and technicians. But what happens if we find a more comprehensive theory in the sense of Sect. 5.1? Is it possible that contradictions can arise between the reality domains of various physical theories? It is not possible, since if $PT_2$ is a more comprehensive theory than $PT_1$, then we can deduce $MT_{\Delta_1}$ from $MT_{\Delta_2}$, so that a theorem in $MT_{\Delta_1}$ is also a theorem in $MT_{\Delta_2}$. Therefore $W_2$ contains $W_1$. The growth of physics also leads to the growth of the reality domain.

As explained in Sect. 6.4 one can take some real numbers $\eta$ in $\mathcal{H}$ as "given." Then we have the possibility to obtain a detailed reality domain, i.e., we can interpret $G(A, \mathcal{H})$ not only as an imprecision but also as a set of various realities for the various "given" values of $\eta$ (see Sect. 6.4). This signifies a more detailed description of $\eta$ that follows directly from the descriptions in Sect. 6.5.

The most interesting problem is the relation between two theories which seem to be different, but which are indeed two forms of the same theory.

# Part II

Examples of Simple Theories

# A

# A Description of the Surface of the Earth, or of a Round Table

This example was used throughout Part I to illustrate some of the concepts introduced.

## The Physical Reality

As a physical system, we take marked spots and distances between marked spots.

## The Basic Language $B_l$

In this context we consider the following:

- Only *one* property concept *"marked spot"* denoted by $\tilde{p}$. In the basic language $B_l$ we will formulate sentences such as 'the object $\bar{a}$ has the property *marked spot*'.
- Only *one* relation concept *"distance relation between objects"* denoted by $\tilde{r}$. In the basic language $B_l$ we will formulate sentences such as 'between the objects $\bar{a}_1, \bar{a}_2$ and the real number $\alpha$ there is the distance relation $\delta(\bar{a}_1, \bar{a}_2, \alpha)$', where the *distance relation* $\delta(\bar{a}_1, \bar{a}_2, \alpha)$ is obtained on the basis of a pre-theory of measurement of distance.

## The Application Domain $A_p$

We decide that the property concept "marked spot" is to be taken as a basic property. This means that we want to describe by the intended theory only such objects which are marked spots. The application domain $A_p$ consists of – marked spots – and – distance relations between marked spots –.

## The Text $\widetilde{A}$

We have a finite collection of sentences, denoted by $\widetilde{A}$, such as

'the object $\bar{a}_1$ has the property marked spot *and* the object $\bar{a}_2$ has the property marked spot *and* between the objects $\bar{a}_1, \bar{a}_2$ and the number $\alpha$ there is the distance relation $\delta(\bar{a}_1, \bar{a}_2, \alpha)$'.

## The Basic Mathematical Theory $MT_{\widehat{\Theta}}$

We add to $MT$ two new constants $\bar{p}$ and $\bar{r}$: $\bar{p}$ is a relation of weight 1 noted $\bar{p}(x)$, and $\bar{r}$ is a relation of weight 3 noted $\bar{r}(x_1, x_2, \alpha)$, where $\alpha$ is a real number $\alpha \in \overline{\mathbb{R}}$ and $\overline{\mathbb{R}}$ is a finite set of real numbers. We have taken $\widetilde{p}$ as a basic property. Therefore, on the basis of the axiom (3.2.1), there is a set $\overline{M}$ with

$$x \in \overline{M} \Leftrightarrow \bar{p}(x).$$

On the basis of the axiom (3.2.2) $\overline{M}$ is a finite set.

(We have introduced $MT_{\widehat{\Theta}}$ only to show that the "collectivizing" axiom is the basis of the following standard form $MT_\Theta$.)

## The Standard Mathematical Theory $MT_\Theta$

We add to $MT$ two new constants $\overline{M}$ and $\bar{s}$. On the basis of the axiom (3.2.3) $\overline{M}$ is a finite set. On the basis of the axiom (3.2.5) we have

$$\bar{s} \subset \overline{M} \times \overline{M} \times \overline{\mathbb{R}}.$$

The mathematization process $B_l$ (cor) $MT_\Theta$, i.e., the transcription of natural sentences formulated in the basic language into formal sentences formulated in the formal language, is given by:

'$\bar{a}$ is a marked spot'                                      (cor) '$\bar{a} \in \overline{M}$',

'the measured distance between $\bar{a}$ and $\bar{b}$ is $\alpha \pm \varepsilon$'    (cor) '$(\bar{a}, \bar{b}, J) \cap \bar{s} \neq \emptyset$'.

Here $J$ is the interval $\alpha - \varepsilon$ to $\alpha + \varepsilon$.

# A  A Description of the Surface of the Earth, or of a Round Table

## The Enrichment of $MT_\Theta$ by $\overline{A}$

We add to $MT_\Theta$ new constants $\overline{a}_1, \overline{a}_2, \ldots$ for some *recorded* marked spots. As "axioms" we add the recorded and, by (cor), transcribed facts, i.e., sentences or relations noted by $\overline{A}$:

$$\overline{a}_i \in \overline{M}, \ldots$$

and

$$(\overline{a}_i, \overline{a}_k, J_{ik}) \cap \overline{s} \neq \emptyset, \ldots$$

Since we have in $MT_\Theta$ not introduced any axiom for the subset $\overline{s}$ of $\overline{M} \times \overline{M} \times \overline{\mathbb{R}}$, the theory $MT_\Theta \overline{A}$ cannot lead to contradictions if we have not made recording errors.

## The Idealized Mathematical Theory $MT_\Delta$

We formulate by $\Delta$ that $\overline{M}, \overline{s}$ is "approximately" a two-dimensional Euclidean geometry.

In $MT_\Theta$ we define the following sets:

$$Q = \mathbb{R} \times \mathbb{R}$$

and $s$ as the set

$$s \subset Q \times Q \times \mathbb{R}$$

of all $(q_1, q_2, d)$ with

$$d(\alpha_1, \alpha_2; \beta_1, \beta_2) = \sqrt{(\alpha_1 - \beta_1)^2 + (\alpha_2 - \beta_2)^2}$$

where $q_1 = (\alpha_1, \beta_1)$ and $q_2 = (\alpha_2, \beta_2)$ with $\alpha_i, \beta_i \in \mathbb{R}$.

We define the following inaccuracy set for $Q$:

$$U \subset Q \times Q, \quad U_{\varepsilon r} = \{(q_1, q_2) \mid d(q_1, q_2) < \varepsilon\} \cup Q_r \times Q_r, \tag{A.1}$$

where

$$Q_r = \{q \mid q \in Q \text{ and } d(q, 0) > r\},$$

and in which 0 is the point $(0, 0)$ in $\mathbb{R} \times \mathbb{R}$.

For $\mathbb{R}_+$ we introduce the inaccuracy set

$$U_{\delta\varrho} = \{(\alpha_1, \alpha_2) \mid |\alpha_1, \alpha_2| < \delta\} \cup \mathbb{R}_+^\rho \times \mathbb{R}_+^\rho, \qquad (A.2)$$

where $\mathbb{R}_+^\rho = \{\alpha \mid \alpha \in \mathbb{R}_+ \wedge \alpha > \rho\}$. We take $\delta < \varepsilon$ and $\varrho > 2r\pi$.

These inaccuracy sets generate in a canonical way also an inaccuracy set $U_s$ in $Q \times Q \times \mathbb{R}_+$. On the basis of the axiom (3.3.5) we have

$(\exists \phi)$

$[\phi : \overline{M} \to Q$ is an injective mapping with $(\phi \overline{M})_U = Q$

$\wedge \; \phi\overline{s} \subset (s)_{U_s}$

$\wedge \; \phi\overline{s}' \subset (s')_{U_s}].$ \hfill (A.3)

If we want to give particular values for the $\varepsilon, r$ we must take different application domains, one for the round table and one for the surface of the earth:

- For the round table we can choose, e.g., $\varepsilon = 0.1$ mm and $r$ essentially greater than the radius of the round table, e.g., ten times the radius of the table;

- For the surface of the earth we can choose, e.g., $\varepsilon = 10$ cm and $r = 10$ km. But we can also choose other values. If we choose a bigger $r$, we must also choose a bigger $\varepsilon$.

## The Enrichment of $MT_\Delta$ by $\overline{A}$

In this case we record – marked spots – and the results of measurements of the – distances – between these marked spots.

We presuppose the knowledge of "how to measure a distance," i.e., the knowledge of a pre-theory of measurements of distances, a pre-theory of the use of such measuring records but without the geometry of the surfaces. These pre-theories show that there are no precise measurements. We say that every measurement has a so-called "error of measurement" describing the imprecision of the measurement. The result of a measurement of the distance between two marked spots $\overline{a}, \overline{b}$ may be given by the real number $\alpha$. This then has to be corrected by saying that the value is not exactly $\alpha$, but can be any number between $\alpha - \varepsilon$ and $\alpha + \varepsilon$, i.e., a number of an interval $J$.

Now we want to describe the use of our theory $MT_\Delta$ for the description of the real application domain of marked spots on earth, and their measured distances.

We see by (A.1) that we have an example of the "second case" of Sect. 3.3.2. The region of "large inaccuracies is given by $Q_r$, and the point $(0,0) \in Q$ is a

# A A Description of the Surface of the Earth, or of a Round Table

"center" of the region $Q'_r$ of "small" inaccuracies. We choose one $\bar{a}$ appearing in $\overline{A}$. The selected $\bar{a}$ may be $\bar{a}_0$ ($\bar{a}_0$ can be, e.g., a particular marked spot at Greenwich in London). We add in (A.3) the condition $\phi\bar{a}_0 = (0,0)$.

Thus we get the axiom

$$(\exists \phi)$$
$$[\phi : \overline{M} \to Q \text{ is an injective mapping with } (\phi \overline{M})_U = Q$$
$$\wedge \ \phi\bar{a}_0 = (0,0)$$
$$\wedge \ \phi\bar{s} \subset (s)_{U_s}$$
$$\wedge \ \phi\bar{s}' \subset (s')_{U_s}]. \tag{A.4}$$

A text $\overline{A}$ consists of relations of the form

$$\bar{a} \in \overline{M} \quad \text{and} \quad (\bar{a}_1, \bar{a}_2, J_{12}) \cap \bar{s} \neq \emptyset$$

which lead, for a $\phi$ satisfying (3.3.12), to relations of the form

$$\phi\bar{a} \in Q \quad \text{and} \quad (\phi\bar{a}_1, \phi\bar{a}_2, J_{12}) \cap (s)_{U_s} \neq \emptyset. \tag{A.5}$$

(Relations for $\bar{s}'$ and $s'$ are not interesting because they are mathematically irrelevant.) If we write $\phi\bar{a} = a$, we get relations of the form

$$\bar{a}_0 = (0,0), \quad \bar{a} \in Q \quad \text{and} \quad (\phi\bar{a}_1, \phi\bar{a}_2, J_{12}) \cap (s)_{U_s} \neq \emptyset, \tag{A.6}$$

which is only interesting for such an $a$ with $a \in Q'_r$, i.e., for the surrounding of $a_0 = (0,0)$.

In this example, a mapping $\phi$ satisfying (A.4) has a very graphic signification: $\phi$ generates a map of the earth printed onto a sheet of graph paper. On this map, the surrounding of $\bar{a}_0$ (Greenwich in London) represents very well the distances between various marked spots, but not so well the marked spots lying very far away from $\bar{a}_0$.

## The Fundamental Domain G

If we consider as an application domain $A_p$ the marked spots on the surface of the earth, from Sect. 3.3.2 we see immediately that a fundamental domain $G$ is given by the surrounding of a particular marked spot (e.g., a marked spot in Greenwich) from which we get by (cor)$\phi$ a "map" with small inaccuracies relative to the distances.

152   A A Description of the Surface of the Earth, or of a Round Table

For the case of the surface of a round table, we have as the fundamental domain $G$ the total surface, i.e., the total application domain $A_p$. The problem with this example is the following: Which part of $\overline{M}$ refers to marked spots?

## The Axiomatic Basis $MT_{\widehat{\Delta}}$

We get an axiomatic basis $MT_{\widehat{\Sigma}}$ characterized by

1. the principal base set $\widehat{M}$,
2. the structure term $\widehat{s}$,
3. with the typification $\widehat{s} \subset \widehat{M} \times \widehat{M} \times \mathbb{R}$,
4. the axiomatic relation $\widehat{P}$, according to (4.3.1),

$(\exists f)\ [\ f : \widehat{M} \to Q = \mathbb{R} \times \mathbb{R}\ \text{is a bijective mapping}$

with $f\widehat{s} = s\ ]$,

where $s \subset Q \times Q \times \mathbb{R}$ is the set of all $(q_1, q_2, d)$

with $d(\alpha_1, \alpha_2; \beta_1, \beta_2) = \sqrt{(\alpha_1 - \beta_1)^2 + (\alpha_2 - \beta_2)^2}$   (A.7)

(see Sect. 3.3.1).

One has to insert in the square brackets that for a $\widehat{U}$ corresponding to $U$ of (A.1),

$f\widehat{U} = U.$

A fixing of the picture of one $\overline{a}_0$ (as described in Sect. 3.3.2) can also be transported to $MT_{\widehat{\Delta}}$. With $\phi(\overline{a}_0) = (0, 0, 0)$ we have only to set (with $\widehat{\phi}_i \widehat{q}_0 = \widehat{a}_0$) $f_i(\widehat{a}_0) = (0, 0, 0)$.

If we take the axiomatic relation $\widehat{P}$ of (A.7), we know that $MT_{\widehat{\Sigma}(\widehat{M},\widehat{s})}$ is a two-dimensional Euclidean geometry. But we want to formulate $\widehat{P}$ in such a way that it immediately says something about $\widehat{s}$, so that we can prove later (A.7) as a theorem. Such a system of axioms $\widehat{P}$ can, e.g., begin with the axioms

1. $\widehat{s}$ determines a mapping $\widehat{M} \times \widehat{M} \to \mathbb{R}_+$, i.e., a real function $d(x_1, x_2) \geq 0$,
2. $d(x_1, x_2) + d(x_2, x_3) \geq d(x_1, x_3)$,

and can be extended by additional axioms of a similar form.

In this way we can find a $\widehat{P}$ which "depends" only on $\widehat{M}$ and $\widehat{s}$, i.e., an axiomatic relation where the existence of terms is postulated only if these terms can be "formulated" by $\widehat{M}$ and $\widehat{s}$ and not by the help of terms formulated

# A A Description of the Surface of the Earth, or of a Round Table

by real numbers (as, e.g., $\mathbb{R} \times \mathbb{R}$). It is not necessary for the formulation of $\widehat{P}$ to "imagine" terms other than $\widehat{M}$ and $\widehat{s}$.

## The Inaccuracy Sets and Uniform Structures

The inaccuracy set $U_{\delta\varrho}$ given in (A.2) is an element of the set $N$ generated by $d(\alpha_1, \alpha_2)$. With this set $N$ the set $\mathbb{R}$ is precompact.

The completion $\widetilde{\mathbb{R}}$ of $\mathbb{R}$ is $\mathbb{R}$ plus two elements; one is in $+\infty$ and the other is in $-\infty$. It is clear that we have to choose out of this $N$ different inaccuracy sets $U$ for different $s(\ldots,\mathbb{R})$.

Also for the set $Q = \mathbb{R} \times \mathbb{R}$ (see Sect. 3.3.1) we can find a metric (which generates a set $N$ of which (3.3.6) is an element) by the following procedure (see Fig. A.1).

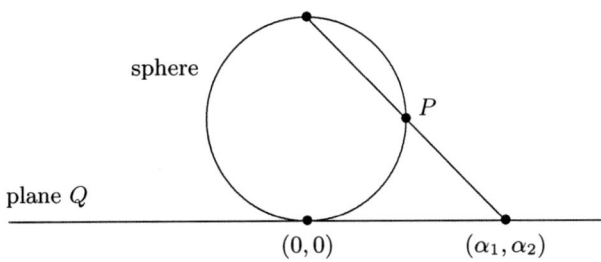

**Fig. A.1.** Stereographic projection

Every point $(\alpha_1, \alpha_2)$ is mapped onto a point $P$ on the surface of a sphere. As a metric on $Q$ we use the distance of the corresponding points on the surface of the sphere. With this metric, $Q$ is precompact. $\widetilde{Q}$ is the set $Q$ plus one point in the "infinity" of $Q$.

But we can also use another metric with another $\widetilde{Q}$, taking the mapping of Fig. A.2.

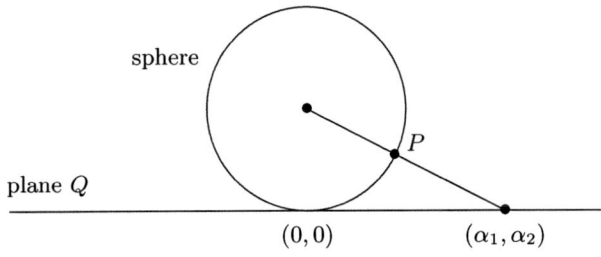

**Fig. A.2.** Gnomonic projection

The set $N$ contains the set (A.1) as an element. For this set $N$, the set $\tilde{Q}$ is the set $Q$ plus different points in "infinity." But for different directions from $(0,0)$ to "infinity" there are also different points.

This example demonstrates that the set $N$ selected is not determined by reality. The set $N$ selected also has to do with our intention to obtain such inaccuracy sets $U$ for which $\Delta$ generates a "good" physical theory.

In our example $Q = \mathbb{R} \times \mathbb{R}$, we can also choose a third uniform structure $N$, namely generated by the distance

$$d(\alpha_1, \alpha_2, \beta_1, \beta_2) = \sqrt{(\alpha_1 - \beta_1)^2 + (\alpha_2 - \beta_2)^2}.$$

With this metric, $Q$ is not precompact! Therefore this set $N$ is not very efficient in *seeking* usable inaccuracy sets; $U$ from (A.1) is an element of this $N$, but this $N$ contains too many elements for finding usable inaccuracy sets.

## The Description of the Reality

If we take as $\overline{A}$ the measured distances between two marked spots, whereby one marked spot is used for the measurement of *one* distance, we cannot have a contradiction to $MT_\Delta$. This shows that these measurements are suitable for making a relevant test of the theory.

One has to try to find a "critical" experiment, i.e., an experiment by which an axiom or a theorem in $PT_\Delta$ can perhaps be refuted. In our Example A (applied to a round table) we can, e.g., perform the following experiment.

According to the pre-theory of measurements of distance we use measuring tapes. We stretch a measuring tape between two marked spots $a$ and $b$; the length of this tape is a measured distance $\delta(a,b)$ of the marked spots $a, b$. (We do not use a ruler, since the definition of a ruler already presumes an Euclidean geometry; we also do not use compasses, since this presumes the definition of a "rigid body." We do not consider constructions using a ruler and compasses.)

We begin the experiment with two marked spots $a, b$. We take a measuring tape, the length of which is greater than the distance $\delta(a, b)$. We fix the ends of this tape onto $a$ and $b$. The tape is not stretched, since the length is greater than $\delta(a, b)$. But we can stretch the tape by joining the middle of the tape to a marked spot $c$, so that the distance $\delta(a, c) = \delta(b, c) = \alpha$, with $2\alpha$ having the length of the tape. But there is also another marked spot $d$, by which the tape is also stretched, so that $\delta(a, d) = \delta(b, d) = \alpha$. We can now measure the distances $\delta(a, b)$ and $\delta(c, d)$; we define $\beta$ and $\gamma$ by $2\beta = \delta(a, b)$ and $2\gamma = \delta(c, d)$. We can now ask whether these measured values $\alpha, \beta, \gamma$ fulfill with a certain inaccuracy the equation $\alpha^2 = \beta^2 + \gamma^2$.

# A  A Description of the Surface of the Earth, or of a Round Table

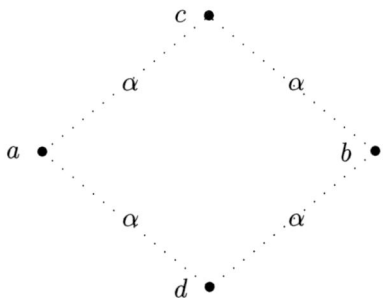

**Fig. A.3.** Experiment with marked spots

If this is the case, we have no contradiction in $MT_\Delta \overline{A}$. If we repeat this experiment a few times, starting with other marked spots $a, b$, and get no contradiction to $MT_\Delta$, we are convinced that our theory is correct, i.e., we trust our theory. We trust it because we do not believe that reality will deceive us.

We can now repeat the same experiment on the surface of a globe (instead of the surface of the earth, since on earth it is more difficult to measure distances, i.e., since we need for the earth a richer theory of measurement of distances.) The result of such experiments is then the following: The equation $\alpha^2 = \beta^2 + \gamma^2$ is only fulfilled with small inaccuracies if the $\alpha, \beta, \gamma$ are small compared to the greatest distance of two marked spots on the surface of the globe. The way out is either to introduce large inaccuracies and as a consequence a smaller fundamental domain $G$ than the total globe, or to go toward a "better" theory by instead of using an Euclidean idealization, we use the idealization of the geometry of the surface of a sphere (see Sect. 3.3.2).

## A Closed Theory

In the case of a round table, if we choose $Q = \mathbb{R} \times \mathbb{R}$ (see Sect. 3.3.1), then this $Q$ is "too large." Indeed, elements in $\overline{M}$, the pictures of which in $Q$ determine much greater distances than the diameter of the real table, cannot be excluded. We can then introduce $Q_s$ as the set of all $(\alpha, \beta) \in \mathbb{R} \times \mathbb{R}$ with the property $\alpha^2 + \beta^2 < r^2$ with $r$ as the measured radius of the table.

The introduction of $Q_s$ for this example is not so complicated, because we can see that "there are" no spots lying outside of the table. In other cases in physics we cannot "survey" the existence or nonexistence of the objects corresponding to the elements of $Q$. In most cases we are in the same position as that of a blind man in front of the table. A blind man can only try to mark spots on the table and to perceive (by feeling), but he does not succeed to mark spots "outwards" from the table. Such experimental features are indications that we have "too large" picture terms $Q_i$ and how we can perhaps find smaller $Q_{i_s}$.

In the case of a round table, $\widehat{\Sigma}$ is given by the set $M$ of points on a circle of radius $r$ (the table has a radius of $r \pm \varepsilon$), a structure $s \subset M \times M \times \mathbb{R}$, and an axiom that $s$ determines a distance $d: M \times M \to \mathbb{R}$, so that $M$ is the set of points in a circle of radius $r$ of a two-dimensional Euclidean space. This theory of the round table is a closed theory. ($\widehat{\Sigma}$, in the mathematical sense, is not an Euclidean geometry!)

## New Concepts

Our task is not only to detect "nonmeasured" realities, but also to detect new realities which we did not know before the introduction of $\widehat{\Sigma}$. Therefore we have to seek for concepts other than those which are described by the base sets $\widehat{M}$ and the structure terms $\widehat{s}$ of the axiomatic basis $\widehat{\Sigma}$.

As a new intrinsic term $E_2$ we introduce the subset $E_2 \subset \mathcal{P}(M)$ of "straight lines" (lying *on* the circle). As $E_1$ we take $E_1 = M$. As new relation $u \subset E_1 \times E_2$, we take the set $u = \{(x,y) \mid x \in y, x \in E_1, y \in E_2\}$ (i.e., "$x$ is a point on the straight line $y$"). To get a new extended structure $MT_{\Delta_{\text{ex}}}$, we add to $\Theta$ the new constant $\overline{E}_2$ with the axiom $\overline{E}_2 \subset \mathcal{P}(\overline{M})$, a relation $\overline{u} \subset \overline{M} \times \overline{E}_2$, and the axiom that the mapping $\phi: \overline{M} \to M$ of the axiom of $\Delta$ fulfills the relations $\phi \overline{E}_2 \subset (E_2)_U$ and $\phi \overline{u} \subset (u)_U$, where $(E_2)_U, (u)_U$ are the "inaccurate" $E_2$ and $u$, i.e., $E_2$ and $u$ enlarged by the inaccuracy sets $U$.

As new concepts in $B_{l_{\text{ex}}}$ we introduce the following correspondences:

- for $\overline{y} \in \overline{E}_2$ : $\overline{y}$ is a "marked straight line;"
- for $(\overline{x}, \overline{y}) \subset \overline{u}$ : the marked spot $\overline{x}$ lies on the marked straight line $\overline{y}$.

The fact that a marked straight line is determined by two marked spots lying on this marked line has to be proved.

## Classifications and Interpretations of the New Concepts

As $MT_{\widehat{\Sigma}}$ we have a circle of a two-dimensional Euclidean geometry described by a set $M$ of points and a distance relation $s(x, y, \eta) \subset M \times M \times \mathbb{R}$ (see Sect. 4.3). As $E_1, E_2$ we introduce

$$E_1 = M, \quad E_2 \text{ the set of straight lines;} \tag{A.8}$$

as relation $u$ we introduce

$$u = \{(y, x) \mid y \in E_1, x \in E_2, y \in x\}; \tag{A.9}$$

# A A Description of the Surface of the Earth, or of a Round Table

and as given recorded $A$ we assume

$$A \;:\; a_1 \in M \;,\; a_2 \in M \;,\; d(a_1, a_2) \sim \eta. \tag{A.10}$$

where $d(a_1, a_2) \sim \eta$ can be written for "the measurement of the distance between the two marked spots $a_1, a_2$ is approximate to $\eta$."

As $\mathcal{H}$ we introduce

$$\begin{aligned}\mathcal{H} \;:\; & x \in (E_2)_U, \\ & a_1 \in E_1 \;,\; a_2 \in E_2, \\ & (a_1, x) \in (u)_U \;,\; (a_2, x) \in (u)_U,\end{aligned} \tag{A.11}$$

that is, the recorded spots $a_1, a_2$ lie approximately on the line $x$.

The set $G(A, \mathcal{H})$ is then the set of all straight lines $x$ for which $a_1$ and $a_2$ lie approximately on these lines. We see that $G(A, \mathcal{H}) \neq \emptyset$ is a theorem. If $d(a_1, a_2)$ is large compared to the inaccuracy $U$, the set $G(A, \mathcal{H})$ is only a neighborhood of one straight line. If $d(a_1, a_2)$ is smaller than $U$, the set $G(A, \mathcal{H})$ contains a total cluster of straight lines around $a_1, a_2$ lying together. So we say that the two spots $a_1, a_2$ record a straight line if they are separated enough.

Later one can state by experiment that a "ruler" can also be used to record a straight line, e.g., by drawing with a ruler and a pencil a straight line onto the round table, which is defined as being an element of $E_2$.

# B

# A Simplified Example of Newton's Mechanics

Newton's mechanics is not a physical theory in our sense, since in general one has to add a concrete form of the so-called "forces;" therefore, Newton's mechanics is a frame theory in the sense of the Sect. 4.9. To make a simplified example of Newton's mechanics we choose a particular situation, as simple as possible, so as not to charge our example with unnecessary mathematical difficulties.

## The Physical Reality

As a physical system we take one material element under the force of a spring. We will include in the application domain $A_p$ experiments with various material elements, but only one spring. Space and time (i.e., the reference systems used to describe the motion of the material elements) will be given by pretheories. We will simplify mathematically the space, i.e., by only *one* coordinate $x$. We contemplate only the motion of one marked spot on the material elements.
Simply:

> The marked spot of the material element is moving on a trajectory $x_i(t)$ which satisfies the differential equation
>
> $$m_i \ddot{x}_i(t) + a x_i(t) = 0. \tag{B.1}$$

Here $m_i$ denotes the mass of the material element $i$ and $a$ the constant of the spring. But we do not know by pre-theories what the "mass" is. Therefore, we cannot take the concept of "mass" into the basic language $B_l$! So what can be taken into this basic language $B_l$?

## The Basic Language $B_l$

Let us recall that facts of the application domain $A_p$ are denoted by natural sentences of the basic language $B_l$ designating propositions of the conceptual system. This means that only facts denoted by natural sentences, using terms that designate concepts belonging to the conceptual system, can be taken into account.

In this example the context is characterized by the following properties.

## Property $\tilde{p}_1$

$\tilde{p}_1$ is the property - to be a material element -; usually one says - to be a mass point -. But the related concept of "mass point" is not very clear since we do not know what a mass is, and the material element is not a mathematical point. A material element is a separate part of a rigid material. The property $\tilde{p}_1$ is taken as one of the basic properties.

It must be possible to recognize this material element as being the same during its motion in the space and time reference system. This reference system will be given by a pre-theory, so that we can give the location of the material element (i.e., of the marked spot on it) at a time by a real number $x$ (we simplify mathematically the space by *one* real number) and the time by a real number $t$. We can take these numbers as elements of a finite set $\overline{\mathbb{R}}$ of real numbers. The so-called "measuring errors," or better, the imprecision of these $x$ and $t$, are given by the measuring methods and the pre-theories.

It is clear that we cannot expect that the $x,t$ of one "experiment" depend on those of another experiment; that is, we must distinguish the various experiments. But what signifies the word 'experiment'? The signification of this word, or the meaning of the related concept "experiment," will give us a second basic property.

## Property $\tilde{p}_2$

As "experiment" we consider the following situation which can also be caused by other circumstances given in nature. An experiment produces a "free" motion of a material element. This motion depends on the circumstances by which the motion was produced; but we will not describe here these circumstances. We are only interested in the produced "free" motion; a motion not influenced by any thing other than the spring. Such free motions may be denoted by the word 'experiment'. $\tilde{p}_2$ is the property that "the motion is an experiment."

(It is clear that one can have difficulties of stating correctly that the motion is "free;" one cannot have observed some influences which show that $\tilde{p}_2$ is not valid. But if we are not sure, we do not take this case in the application domain $A_p$.)

(A second remark has to be made! A measurement of $x,t$ is not possible without the influence of other physical realities on the material element, for example light. The assumptions of a "free" motion and of a measurement of $x,t$ are contradictory! But from experiment we have that for macroscopic material elements, the influence of the measurements can be reduced; for example, we do not observe that the light is influencing the motion, except if one constructs very precise systems. If the influence of a measurement is small, we speak of weak measurements, otherwise, of hard measurements. If, e.g., the material elements make a hole in another piece of material, this hole is a measurement of the location of the element at that time when the hole was made.)

The condition of a "free" motion is valid only for a finite time interval, and we have to describe this in the theory. Thus we introduce into the basic language $B_l$ the relation

'$\widetilde{r}_1(b, i, x, t)$'

corresponding to

'In the experiment $b$ the spot on the material element $i$ has the (imprecise) position $x$ at the (imprecise) time $t$'.

For the description of the time interval, where the motion of the material element is "free," we do not introduce here a relation; we will use this feature later in choosing the inaccuracy sets. We introduce only as a second relation

'$\widetilde{r}_2(b, i)$'

corresponding to

'In the experiment $b$ the material element $i$ is moving'.

## The Basic Mathematical Theory $MT_{\widehat{\Theta}}$

To $MT$ four constants are added: $\overline{p}_1, \overline{p}_2, \overline{r}_1, \overline{r}_2$, where $\overline{p}_1, \overline{p}_2$ are relations of weight 1, $\overline{r}_1$ is a relation of weight 4, and $\overline{r}_2$ is a relation of weight 2. The mathematization process (cor) is simple.

## The Standard Mathematical Theory $MT_{\Theta}$

To $MT$ four constants are added: two sets $\overline{M}_1, \overline{M}_2$ and two sets $\overline{s}_1 \subset \overline{M}_2 \times \overline{M}_1 \times \mathbb{R} \times \mathbb{R}$ and $\overline{s}_2 \subset \overline{M}_2 \times \overline{M}_1$. The mathematization process (cor) is simple.

## The Idealized Mathematical Theory $MT_{\Delta}$

To $MT$ three constants are added: two sets $M_1$ and $M_2$ and a set $m$ with the axiom $m \subset M_1 \times \mathbb{R}_+$ such that $m$ determines a mapping $m : xsM_1 \to \mathbb{R}_+$. For $i \in M_1$ we write simply $m_i$ instead of $m(i)$.

In addition, to $MT$ two constants are added: two sets $s_2$ and $s_3$ with the axioms $s_2 \subset M_2 \times M_1$ such that $s_2$ determines a mapping $\sigma : M_2 \to M_1$, and $s_3 \subset M_2 \times \mathbb{R} \times \mathbb{R}$ such that $s_3$ determines a mapping $\varrho : M_2 \to \mathbb{R} \times \mathbb{R}$.

(To simplify the relations we will set in (B.1) $a = 1$. We can do this because we consider *only* one spring and the number $a$ determines only the units.)

To formulate $\Delta$, we use as picture terms for $\overline{M}_1, \overline{M}_2, \overline{s}_2$ the sets $M_1, M_2, s_2$, and as a picture term for $\overline{s}_1$ we use $s_1 \subset M_2 \times M_1 \times \mathbb{R} \times \mathbb{R}$:

$$s_1 = \Big\{(b, i, x, t) \;\Big|\; i = \sigma(b),$$
$$x = x(t) \text{ with } m_i \ddot{x}_i(t) + x_i(t) = 0,$$
$$x(0) = \alpha_1,\; \ddot{x}(0) = \alpha_2 \text{ with } (\alpha_1, \alpha_2) = \varrho(b)\Big\}.$$

Because of $i = \sigma(b)$, it follows that

$$s_1 \subset s_2 \times \mathbb{R} \times \mathbb{R}.$$

Until this point we have only given sets and axioms in $MT$, i.e., we have only given a mathematical game without any connection to the reality. Only $MT_\Theta$ is related to the reality.

The meaning of $M_2, s_1, s_2$, i.e., the sense of the conceptual entities and their reference to reality, is that we will formulate by $\Delta$ that $\overline{M}_1, \overline{M}_2, \overline{s}_1, \overline{s}_2$ (of $\Theta$) shall be "similar" to the $M_1, M_2, s_1, s_2$ (of $MT$) – similar but not "equal," i.e., not necessarily isomorphic. For this purpose we enrich $\Theta$ to $\Delta$ by using $M_1, M_2, s_1, s_2$ as auxiliary terms for $\Delta$.

To formulate the "similarity," we add at first new constants $\phi_1, \phi_2$ to $MT_\Theta$ with the axioms: $\phi_1, \phi_2$ are mappings

$$\phi_1 : \overline{M}_1 \to M_1, \quad \phi_2 : \overline{M}_2 \to M_2.$$

We especially take the axiom that $\phi_1$ and $\phi_2$ are *bijective* and we postulate the axiom that

$$\phi \overline{s}_2 = s_2.$$

With the identical mapping $\overline{\mathbb{R}} \to \mathbb{R}$ we get also a mapping

$$\phi : \overline{M}_2 \times \overline{M}_1 \times \overline{\mathbb{R}} \times \overline{\mathbb{R}} \to M_2 \times M_1 \times \mathbb{R} \times \mathbb{R}.$$

Because of the axiom $\phi \overline{s}_2 = s_2$, we get that $\phi$ is also a bijection of $\overline{s}_2 \to s_2$. A physicist knows that the strong axiom $\phi \overline{s}_1 = s_1$ leads to contradictions with experiments. But we postulate at least the axiom that

$$\phi \overline{s}_1 \subset s_2 \times \mathbb{R} \times \mathbb{R}.$$

## B  A Simplified Example of Newton's Mechanics

From this follows that

$$\bar{s}_1 \subset \bar{s}_2 \times \overline{\mathbb{R}} \times \overline{\mathbb{R}}.$$

Since $\phi \bar{s}_1 = s_1$ is not possible, we have to choose an inaccuracy set $U$ for $s_2 \times \mathbb{R} \times \mathbb{R}$, so that we can postulate the axiom that

$$\phi \bar{s}_1 \subset (s_1)_U, \quad \phi \bar{s}'_1 \subset (s'_1)_U.$$

Because of the bijection $\phi : \bar{s}_2 \to s_2$, the inaccuracy set $U$ is only related to $\mathbb{R} \times \mathbb{R}$. This means that $U$ is given by diagonal elements of $s_2 \times s_2$ and by an inaccuracy set $U_{(b,i)}$ for $\mathbb{R} \times \mathbb{R}$ for every $(b, i) \in s_2$.

The choice of $U_{(b,i)}$ is based on physical experiments. Therefore, we have to discuss physical experiments, i.e., $U_{(b,i)}$ cannot be deduced from intuition.

At first we exclude from our theory all the cases where we expect great differences between $\phi \bar{s}_1$ and $s_1$, i.e., we pass from the application domain $A_p$ to a smaller region of applications, to the so-called fundamental domain $G$ of the theory.

We have already mentioned above that we have to exclude all the times when the system is not "isolated." To every experiment $b$ belongs a finite time interval, where the system is isolated. As this time interval we take the time between 0 and (for $b \in M_2$) $T_b$, where we have chosen the time $t = 0$ as the beginning of this interval. The time $T_b$ is determined by influences from the surroundings of the system to be described.

But there are two other "times" determined by the system itself, which make the description of the motion by the theory seem incorrect:

To every $b \in M_2$ there is a time $T_c$ where the deviation between the theory and the reality will be greater and greater, since there are internal processes in the spring which are not described by the force of the form $(-x)$, e.g., internal "frictional losses" in the spring.

If the material element moves to great values of $x$, the description by the theory is also wrong since in these cases the force of the form $(-x)$ is not a good description, so that after a certain time, where $|x|$ is too great, the description of the trajectory by the theory will be wrong:

To every $b \in M_2$ there is (by $\varrho(b)$) a time $T_d$ which (for the trajectory $x(t)$) will be at first greater than a given $X$.

To exclude all of the times $t \geq T_b$, $t \geq T_c$, $t \geq T_d$ from the application domain $A_p$ of the theory, i.e., to describe the fundamental domain $G$ of the theory, one can follow several methods. The simplest method is to take $s_1$ only for such $t$ with $t \geq T_b$, $t \geq T_c$, $t \geq T_d$ and to compare $s_1$ with $\phi \bar{s}_1$, also only for these values of $t$. For the problem of the comparison of various theories, it is mathematically more suitable to describe this exclusion of times by introducing suitable inaccuracy sets $U_{(b,i)}$.

# B A Simplified Example of Newton's Mechanics

Let $R_{(b,i)}$ be the set of all $t$ for which $t \geq T_b$, $t \geq T_c$, $t \geq T_d$. We choose as $U_{(b,i)}$ the set

$$U_{(b,i)} = \left\{(x_1, t_1), (x_2, t_2) \mid |x_1 - x_2| < \varepsilon, |t_1 - t_2| < \eta\right\}$$
$$\cup \left[(R \times R_{(b,i)}) \times (R \times R_{(b,i)})\right].$$

(It is clear that one could also choose smaller $U_{(b,i)}$, but we do not wish to do so.)

The small numbers $\varepsilon$ and $\eta$ have nothing to do with the "measuring errors" of spatial deviation $x$ and time $t$! These so-called "errors" originate from the pre-theories and have to be described by "intervals" instead of by exact numbers! If the errors are large, one can also get a good theory for $\varepsilon = 0$, $\eta = 0$. The numbers $\varepsilon$ and $\eta$ describe the fact that the reality cannot be described precisely by a trajectory $x(t)$ which is, in addition, also differentiable; a mathematical property that has no physical sense.

This example B will also serve to illustrate the investigations of Chap. 6. In $MT$ the so-called mass $m_i$ of the material elements was introduced. Since the concept of "mass" has no meaning in the context of the application domain $A_p$, it is a typical example of an "imagined" concept which refers to an "imagined" reality, i.e., a fairy tale. The fact that we have a usable theory is not enough to show that the $m_i$ are real; what is of importance is the noncontradiction with experiments. In Chap. 6 we described that the real (imprecise) numbers $m_i$ do indeed describe a reality.

Another question is the so-called determination of the trajectory by the "initial values." This is the case for idealized trajectories $x(t)$, but is it also the case like in real life? In Chap. 6 we saw that a nondeterministic evaluation can be described imprecisely by deterministic idealizations, e.g., if the idealized evaluations are unstable, i.e., if small deviations at the beginning can lead to too large deviations later on.

The mathematization of the imprecision between the idealizations compared to reality seems to be complicated. But we have only introduced a mathematical description of those procedures which an experimental physicist has to introduce in order to compare his experimental results with the "idealized" theory. He does this by a mixture of everyday language and mathematical symbols (e.g., numbers). Our mathematical analysis will not replace the work of the experimental physicist. Also, a mathematician will not provide in any case all of the steps of the proof, in the sense of the analysis of Sect. 2.2.1.

Our intention is only to show in principle that there is a difference between the reality described in $MT_\Theta$ and the "idealized" description in $MT$. This will be of great importance if we compare different physical theories (see Chap. 5).

This example B could perhaps serve as an argument against our description of the results of experiments by only finitely many relations in $\widetilde{A}$ and $\overline{A}$. One could perhaps mean that a "trajectory" $x(t)$ could be measured continuously;

but this is not the case, since we have no possibility to determine continuously many "time-points." The modern technical method used to record all measurements digitally shows very clearly that we always have only finitely many relations in $\widetilde{A}$. Nevertheless, the number of these relations may be very large.

# C

# The Structure of the Human Species

It will be shown by this example that the method of mathematization can also be applied to nonphysical structures. We call a structure a nonphysical structure if the concepts used in the basic language $B_l$ are "nonphysical" concepts. Physics is determined not only by the applied method, but also by used concepts. Many concepts in physics are defined by pre-theories. But all theories have to start from "first" theories which do not use pre-theories, and the concepts used in these "first" theories are those which determine "physics." The number of these "first" concepts have become smaller and smaller during the development of physics. Today, concepts such as, e.g., "warm" and "cold," "red" and "blue," "strong" and "weak" are no longer used as "first" concepts.

In this book we do not have to deal with the problem of what are today considered the "first" concepts from which all other physical concepts can be defined by theories.

For the structure of human species we introduce in $B_l$ only terms which designate the following concepts:

We introduce the property concept

"to be human."

We decide that this concept is a "basic property" concept, i.e., the application domain $A_p$ is only composed of human beings.

We introduce the quantitative concept

"number of the year."

We consider that this concept is known from a pre-theory, i.e., we consider the "the year $n$" as known, where $n$ is a natural number, e.g., "the year 1983."

As nonbasic property concepts we introduce

"to be female,"

"to be male."

As 2-ary relation concepts we introduce

"to be the mother of,"

"to be the father of,"

"to be born in the year $n$,"

"to have died in the year $n$."

As it is not difficult to formulate $MT_{\widehat{\Theta}}$, we will immediately describe $MT_{\Theta}$.

## The Standard Mathematical Theory $MT_{\Theta}$

We introduce a set $\overline{M}$ as a constant and postulate the axiom that $\overline{M}$ is a finite set. We introduce a set $\overline{Z}$ as a set of natural numbers $n$ with $-N < n < N$, where $N$ is a very large number, e.g., $N = 10^{10}$. We introduce as additional constants the following relations as subsets:

$\overline{\mu} \subset \overline{M}$,

$\overline{\varphi} \subset \overline{M}$,

$\overline{m} \subset \overline{M} \times \overline{M}$,

$\overline{f} \subset \overline{M} \times \overline{M}$,

$\overline{l} \subset \overline{M} \times \overline{Z}$,

$\overline{d} \subset \overline{M} \times \overline{Z}$.

The mathematization process $B_l$ (cor) $MT_{\Theta}$ (i.e., the transcription of natural sentences formulated in the basic language $B_l$ into formal sentences formulated in the formal language $MT_{\Theta}$) is given by

'$\overline{a}$ is female'         (cor)   '$\overline{a} \in \overline{\varphi}$',

'$\overline{a}$ is male'           (cor)   '$\overline{a} \in \overline{\mu}$',

'$\overline{a}$ is the mother of $\overline{c}$'   (cor)   '$(\overline{a}, \overline{c}) \in \overline{m}$',

'$\overline{b}$ is the father of $\overline{c}$'   (cor)   '$(\overline{b}, \overline{c}) \in \overline{f}$',

'$\overline{a}$ was born in $n$'   (cor)   '$(\overline{a}, n) \in \overline{l}$',

'$\overline{a}$ has died in $n$'   (cor)   '$(\overline{a}, n) \in \overline{d}$'.

## The Idealized Mathematical Theory $MT_\Delta$

We now make a particular selection of the theory $MT$. We introduce in $MT$ the following constants: a set $M$ and the subsets

$\mu \subset M$,

$\varphi \subset M$,

$m \subset M \times M$,

$f \subset M \times M$,

$l \subset M \times Z$,

$d \subset M \times Z$,

where $Z$ is the set of all natural numbers. We postulate the following axioms in $MT$:

1. $M$ is countable,
2. $\mu \cap \varphi = \emptyset$, $\mu \cup \varphi = M$,
3. $l$ and $d$ are mappings $M \to Z$,
4. $(a, n_1) \in l$ and $(a, n_2) \in d \Rightarrow n_1 \leq n_2$,
5. $(a_1, c) \in m$ and $(a_2, c) \in m \Rightarrow a_1 = a_2$ and $a_1 \neq c$,
6. $(a_1, c) \in f$ and $(a_2, c) \in f \Rightarrow a_1 = a_2$ and $a_1 \neq c$,
7. $(a, c) \in m \Rightarrow a \in \varphi$,
8. $(a, c) \in f \Rightarrow a \in \mu$,
9. $\Big[ (a, c) \in m \text{ or } (a, c) \in f \Big]$ and $(a, n_1) \in l$ and $(c, n_2) \in l$
   $\Rightarrow n_1 + 10 < n_2$,
   $\Big[ (a, c) \in m \text{ or } (a, c) \in f \Big]$ and $(a, n_1) \in d$ and $(c, n_2) \in l$
   $\Rightarrow n_2 < n_1 + 1$,
10. $c \in M \Rightarrow \exists a \Big[ (a, c) \in m \Big]$,
    $c \in M \Rightarrow \exists b \Big[ (b, c) \in f \Big]$.

With this $MT$ we define $\Delta$.

We introduce in $MT_\Theta$ a constant $\phi \subset \overline{M} \times M$, and add an axiom that $\phi$ is an injective mapping $\phi : \overline{M} \to M$.

The mapping $\phi$ cannot be surjective, since $\overline{M}$ is finite and $M$ must be infinite because of the axiom (10). The picture $\{M, \mu, \ldots, d\}$ is an idealized picture, since it does not contain the possibility that human beings have developed from nonhuman beings. As we do not wish to wait until we know the exact description of the development of the human species, we take the picture described above as an "imprecise" picture, and describe the imprecision by the following inaccuracy set for $M$.

We take a year $n_0$ from which we know that the human species was already existent, and define in $M \times M$ the following set $U$ as an inaccuracy set

$$U = \delta \cup M_0 \times M_0,$$

where $\delta$ is the diagonal in $M \times M$ and

$$M_0 = \left\{ x \mid (x, n) \in l \text{ for an } n < n_0 \right\}.$$

We add the axiom that $\phi \overline{M}$ is $U$-dense in $M$ and the axioms

$$\phi\overline{\mu} \subset (\mu)_U, \quad \phi\overline{\mu}' \subset (\mu')_U;$$
$$\phi\overline{\varphi} \subset (\varphi)_U, \quad \phi\overline{\varphi}' \subset (\varphi')_U;$$
$$\phi\overline{m} \subset (m)_U, \quad \phi\overline{m}' \subset (m')_U;$$
$$\phi\overline{f} \subset (f)_U, \quad \phi\overline{f}' \subset (f')_U;$$
$$\phi\overline{l} \subset (l)_U, \quad \phi\overline{l}' \subset (l')_U;$$
$$\phi\overline{d} \subset (d)_U, \quad \phi\overline{d}' \subset (d')_U.$$

Thus we get a theory without contradiction to the facts, which can be described in the basic language $B_l$.

This example of a theory is also suitable as an example for the description of "measuring errors."

If we have a human being $b$ who was born more than ten thousand years ago, it is not possible to give the exact year of his birth, i.e., we can only say that $(b, n) \in l$ is valid for an $n$ of an interval $J$. The years of this interval may be $n_1, n_2 = n_1 + 1, \ldots, n_p = n_p + 1$. Then we can only say that $(b, n_1) \in l$, or $(b, n_2) \in l$, or ..., or $(b, n_p) \in l$ is valid, i.e.,

$$(\exists n)\left\{n \in J \text{ and } (b, n) \in l\right\}.$$

The "error" interval $J$ can be *changed* by different measurements. But this has nothing to do with the inaccuracy set $U$!

In this example we can measure by pre-theories the "duration of the life of $b$" with much smaller "errors" than the "year of birth." Here the "duration of the life of $b$" is *defined* as the number $n_2 - n_1$ with $(b, n_1) \in l$ and $(b, n_2) \in d$.

# References

1. G. Ludwig: *Die Grundstrukturen einer physikalischen Theorie*, 2nd edn (Springer-Verlag, Berlin Heidelberg New York, 1990). French translation by G. Thurler: *Les structures de base d'une théorie physique* (Springer-Verlag, Berlin Heidelberg New York, 1990)
2. G. Ludwig: *Einführung in die Grundlagen der theoretischen Physik*, 4 vols. (Vieweg, Braunschweig, 1974–1978)
3. E. Scheibe: *Die Reduktion physikalischer Theorien. Teil I: Grundlagen und elementare Theorie* (Springer-Verlag, Berlin Heidelberg New York, 1997)
4. E. Scheibe: *Die Reduktion physikalischer Theorien. Teil II: Inkommensurabilität und Grenzfallreduktion* (Springer-Verlag, Berlin Heidelberg New York, 1999)
5. E. Scheibe: *Between Rationalism and Empiricism: Selected Papers in the Philosophy of Physics, Chapter III Reconstruction*, Ed. by B. Falkenburg (Springer-Verlag, Berlin Heidelberg New York, 2001)
6. N. Bourbaki: *Elements of Mathematics. Theory of Set* (Springer-Verlag, Berlin Heidelberg New York, 1st edition 1968/2nd printing 2004)
7. W. Weidlich, G. Haag: Concepts and models of a quantitative sociology. In: *Series of Synergetics*, Vol. 14 (Springer-Verlag, Berlin Heidelberg New York, 1983)
8. G. Ludwig: *An Axiomatic Basis for Quantum Mechanics*, 2 vols. (Springer-Verlag, Berlin Heidelberg New York, 1986, 1987)
9. N. Bourbaki: *Elements of Mathematics. General Topology*, Chapters 1–4 (Springer-Verlag, Berlin Heidelberg New York, 1st edition 1989/2nd printing 1998)
10. N. Bourbaki: *Elements of Mathematics. General Topology*. Chapters 5–10 (Springer-Verlag, Berlin Heidelberg New York, 1st edition 1989/2nd printing 1998)
11. H.-J. Schmidt: *Axiomatic Characterization of Physical Geometry*, Lecture Notes in Physics, Vol. 111 (Springer-Verlag, Berlin Heidelberg New York, 1979)
12. P. Janich: Protophysik. In: *Handbuch wissenschaftstheoretischer Begriffe*, ed. by J. Speck (Hrsg.) (Vandenhoeck & Ruprecht, Göttingen, 1980)
13. P. Janich: *Die Protophysik der Zeit* (Suhrkamp, Frankfurt/Main, 1980)
14. G. Ludwig: The Relations between various Spacetime Theories in *Semantical Aspect of Spacetime Theories*, ed. by U. Majer, H.-J. Schmidt (BI-Wissenschaftsverlag, Mannheim, 1994)

# List of Symbols

$W$, 11, 14, 144
$PT$ (also written $PT_\nu$), 12
$A_p$ (also written $A_{p_\nu}$), 12, 44
$\varrho$, 12
$G$ (also written $G_\nu$), 13, 60
$W_\nu$, 14, 144
$MT$, 17
$\vee$, 18
$\neg$, 18
$\tau$, 18
$\Rightarrow$, 18
$\tau_x(B)$, 19
$\Leftrightarrow$, 23
$(\exists x)R$, 24
$(\forall x)R$, 24
$=$, 25
$\neq$, 26
$\in$, 27
$\subset$, 27
$\notin$, 27
$\not\subset$, 27
$\mathrm{Coll}_x R$, 28
$\mathcal{E}_x(R)$, 28
$\{x \mid R(x)\}$, 28
$\{x, y\}$, 29
$(x, y)$, 29
$\mathcal{P}$, 30
$\times$, 30
$S(E_1, \ldots, E_n)$, 30
$\langle f_1, \ldots, f_n \rangle^S$, 30
$\hookrightarrow$, 34
$B_l$, 34
$B_{l_{\mathrm{ex}}}$, 40, 132

$J$, 44
$\widetilde{A}$, 45
$\overline{\mathbb{R}}$, 45
(cor), 46
$\widehat{\Theta}$, 47
$MT_{\widehat{\Theta}}$, 47
$\Theta$, 48
$MT_\Theta$, 48
$\overline{M}_0$, 48
$\overline{s}$, 48
$B_l(\mathrm{cor})MT_\Theta$, 49
$\overline{M}_i$, 49
$\overline{A}$, 50, 134
$MT_\Theta \overline{A}$, 50
$\Delta$, 53
$MT_\Delta$, 53
$Q_i$, 53
$s$, 53
$U_i$, 53
$U_s$, 53
$\phi_i : \overline{M}_i \to Q_i$, 54
$P_\Delta$, 55
$\overline{s}'$, 55
$s'$, 55
$MT_\Delta \overline{A}$, 56
$Q_i$, 73
$s_\nu$, 73
$MT_{\Sigma(Q_i, s_\nu)}$, 73
$(MT_\Sigma)_\Delta$, 74
$\widehat{M}_i$, 74
$\widehat{s}_\nu$, 74
$\widehat{\Sigma}$, 74
$MT_{\widehat{\Sigma}}$, 74

## List of Symbols

$\widehat{\Delta}$, 74
$MT_{\widehat{\Delta}}$, 74
$P_{\widehat{\Delta}}$ (also write $\widehat{P}$), 75
$N$, 85
$\Delta_U$, 85
$MT_{\Delta_U}$, 85
$MT_{\Delta_U}\overline{A}$, 85
$MT_{\widehat{\Delta}}\mathcal{H}$, 96
$\Delta_{\text{ex}}$, 113
$MT_{\Delta_{\text{ex}}}$, 113
$PT_{\text{ex}}$, 113
$PT_{\text{ex}} \succ PT$, 114
$B_{l_i}$, 114
$PT_\beta \succ PT_\alpha$, 116
$\Delta_{\text{appr}}$, 117
$PT_{\text{appr}}$, 117
$W_o(\overline{A})$, 124
$\overline{A}_{\max}$, 124
$W_o(\overline{A}_{\max})$, 124
$MT\overline{A}$, 125
$Q_{i_s}$, 125
$PT_s$, 125
$\Sigma_s$, 126
$E_k$, 131

$u_\mu$, 131
$\Sigma_{\text{new}}$, 131
$F$, 132
$T(\widehat{M}_1, \ldots, \mathbb{R})$, 132
$\overline{F}$, 132
$T(\overline{M}_1, \ldots, \overline{\mathbb{R}})$, 132
$\phi : \overline{F} \to T(\widehat{M}_1, \ldots, \mathbb{R})$, 132
$\phi_i : \overline{M}_i \to \widehat{M}_i$, 132
$F_U$, 132
$\overline{E}_k$, 132
$\overline{u}_\mu$, 132
$\widetilde{A}_{\text{ex}}$, 133
$\overline{\mathcal{H}}$, 134
$A$, 136
$\mathcal{H}$, 136
$MT_{\widehat{\Sigma}}A$, 137
$G(\widetilde{A}, \mathcal{H})$, 138
$A_h$, 138
$\mathcal{H}_h$, 138
$MT_{\widehat{\Sigma}}A_h$, 138
$G_h(\widetilde{A}_h, \mathcal{H}_h)$, 138
$G_h(A, \mathcal{H})$, 138
$\overline{G}(\overline{A}, \overline{\mathcal{H}})$, 141
$\widetilde{A}_{\text{tot}}$, 144

# Index

application domain of a $PT$   12, 44
axiom   19
  collectivizing   28, 47
  explicit   19
  finite set   48
  implicit   20
axiomatic basis   73
  simple   77
axiomatic relation   64
  physically interpretable   77
axiomatic rule   20

basic language   34
  extended   40, 132
  initial   116
  semantics of the   39
  syntax of the   38
basic property   40

canonical extension of mappings   30

echelon   30
  construction   30
    scheme   30, 64
experiment
  hypothetical   97

fact   11
  directly recordable   11
  indirectly recordable   11
  not stated   45
  stated   45
finiteness of physics   52
fundamental domain of a $PT$   13, 61

hypotheses
  interpretation of   143
  mathematical classification of   143
  physical classification of   143

idealization process   53
imprecise mapping   54
indirect measurement   134
  inaccuracy set of   138
    classification of   140
      mathematical   140
      physical   141
    interpretation of   141

law of nature
  idealized   75
  idealized pure   77
  pure   76
logics   21

mathematical structure   64
Mathematical Theory
  the basic   46
  the standard   48
mathematical theory   17
  constant of the   19
  the idealized   53
    enriched by $\overline{A}$   57
  the standard
    enriched by $\overline{A}$   50
mathematization process   46
Measurement
  indirect   138
measurement

178    Index

  error of  44, 51
network of physical theories  120
norm  99
now  125

object  11
  hypothetical  139
  possible  135
  property of  11
  relation between  11
physical reality  12
physical system  132
physical theory  12
  application domain of a  12, 44
  approximation  119
  closed  131
  extended  116
  fundamental domain of a  13, 61
  new concept in a  135
  new word in a  136
  reality domain of a  14, 148
  richer than another  116
  skeleton  102
pre-theory  13, 114
proof  19
property  11

reality  11
  possible  131, 135
    new  136
  structure of  11
reality domain of
  a $PT$  14, 148
  all $PT$'s  14, 148
recording process  34
recording rule  45
relation
  between objects  11
  collectivizing  28
  empirically allowed  98
  empirically deductible  98
  empirically refutable  98
  hypothetical  138
  possible  135
  transportable  64

semantic compositionality  41
semantic relation  39
  of denotation  43
  of designation  41
  of reference  42
  of representation  42
sentence
  formal  47
    compound  50
      hypothetical  134
        first kind  138
  natural  34
    compound  38, 45
      extended  135
    negation of a  38
set  27
  auxiliary base  65
  idealization of finite  55
  idealized picture  54
  inaccuracy  54, 85
    possible  86
    usable  87
  physical  54
  principal base  64
  theory of  27
sign  18
  equality  23
  logical  18
  relational  18
  substantific  18
simple axiomatic basis  77
species of structures:  64
  $\Delta$, idealized  66
  $\Delta_{\text{appr}}$, approximation  119
  $\Delta_{\text{ex}}$, extended  115
  $\Sigma$, basic  64
  $\Sigma_{\text{new}}$, related to new concepts  135
  equally rich  66, 108
  equivalent  66
  poorer  66
  procedure of deduction of a  68
  representation of a  71
  richer  66
  theory of the  65
species of uniform structures  86

term
  intrinsic  69
  picture  73
  structure  64
theorem  20
truth of a proposition  44
typification  64

world formula  121

Printed in the United States
52150LVS00002B/22-30